Topics on Oil and Gas

Edited by Ali Ismet Kanli and Tye Ching Thian

Published in London, United Kingdom

Topics on Oil and Gas
http://dx.doi.org/10.5772/intechopen.102202
Edited by Ali Ismet Kanli and Tye Ching Thian

Contributors
Manaf Rizvan Manafov, Gudrat Isfandiyar Kelbaliyev, Dilgam Babir Tagiyev, Ramin Tahmasebi-Boldaji, Jafar Ramezanzadeh, Hossein Moradi, Abdelaziz Khlaifat, Sherif Fakher, Yasir Ali, Yasir Yousif, Alcir de Faro Orlando

First published in London, United Kingdom, 2023 by IntechOpen
IntechOpen is the global imprint of INTECHOPEN LIMITED, registered in England and Wales, registration number: 11086078, 5 Princes Gate Court, London, SW7 2QJ, United Kingdom

British Library Cataloguing-in-Publication Data
A catalogue record for this book is available from the British Library

Additional hard and PDF copies can be obtained from orders@intechopen.com

Topics on Oil and Gas
Edited by Ali Ismet Kanli and Tye Ching Thian
p. cm.
Print ISBN 978-1-83969-662-6
Online ISBN 978-1-83969-663-3
eBook (PDF) ISBN 978-1-83969-664-0

For EU product safety concerns:
IN TECH d.o.o., Prolaz Marije Krucifikse Kozulić 3, 51000 Rijeka, Croatia,
info@intechopen.com or visit our website at intechopen.com.

Meet the editors

Prof. Dr. Ali Ismet Kanlı received his undergraduate degree from the Faculty of Engineering, Department of Geophysical Engineering, Istanbul University, in 1989. He obtained an MSc and Ph.D. from Istanbul University, Institute of Science, in 1994 and 1998, respectively. He began his academic career as a research assistant at the Faculty of Engineering, Department of Geophysical Engineering, Division of Applied Geophysics, Istanbul University in 1992. He became an assistant professor in 2001, an associate professor in 2010, and a professor in 2016. Dr. Kanlı is the head of the Applied Geophysics Division, Geophysical Engineering Department, Istanbul University. He has carried out and directed many international and national projects and has several national and international scientific publications to his credit. He is an editorial board member and reviewer for numerous journals. He has also served as a referee for many international and national projects. He is a member of three national and eight international scientific associations. Dr. Kanlı's scientific interests include applied and near-surface geophysics, engineering and environmental geophysics, engineering seismology, exploration seismology, structural geophysics, earthquake engineering, geotechnical geophysics, borehole geophysics and well logging, alternative energy, and geothermal exploration.

Dr. Tye Ching Thian is an associate professor at the School of Chemical Engineering, Universiti Sains Malaysia. She received her doctoral degree at the University of British Columbia, Canada. She works in the field of chemical reaction engineering and catalysis. She has been involved in projects to improve catalysis activities, system efficiency, and product quality via different upgrading and treating paths related to petroleum and unconventional oil processing, such as heavy oil, used motor oil, and renewable resources like vegetable or plant oil. She previously investigated asphalt characteristic enhancement in road pavement technology. Dr. Thian serves as a reviewer and editorial board member for international and national refereed journals, scientific proceedings, and international and national grants.

Contents

Preface

An oil and gas reservoir is a formation of rock in which oil and natural gas have accumulated. The oil and gas are collected in small, connected pore spaces of rock and are trapped within the reservoir by adjacent and overlying, impermeable layers of rock. Conventional hydrocarbon reservoirs consist of three main parts: the source rock that contains the kerogen that the oil and gas forms from; the reservoir rock, which is the porous, permeable rock layer or layers that hold the oil and gas; and the cap rock, which seals the top and sides so that the hydrocarbons are trapped in the reservoir, while water often seals the bottom. Reservoirs containing only free gas are termed gas reservoirs. Such a reservoir contains a mixture of hydrocarbons, which exists wholly in the gaseous state. With the ongoing energy needs of the world, the importance of oil and natural gas is increasing tremendously in the global market.

There are different types of crude oils with different properties, such as light crude, heavy crude, and extra heavy crude oils, which are categorized according to the American Petroleum Institute (API) degree of the crude. Normally, the heavier the crude, the higher its asphalt content. In addition, crude oils are also termed sweet or sour based on their sulfur content. In a refinery, crude oil processing mainly involves distillation, cracking, hydrocracking, hydrotreating, and blending. Depending on the type of crude used as feedstock, different processes are involved, with the main purpose being to convert as much of the barrel of crude oil into transportation fuels. Petroleum processing technology is rather mature. Nevertheless, it continues to develop in consideration of all types of sustainability-related challenges.

This book provides comprehensive information on the oil and gas industry, in addition to highlighting technological developments in the field. It is a useful resource for geoscientists, petroleum and reservoir engineers, and chemical engineers.

Ali Ismet Kanlı, Ph.D.
Professor of Geophysics,
Faculty of Engineering,
Department of Geophysical Engineering,
Istanbul University-Cerrahpasa,
Istanbul, Turkey

Tye Ching Thian, Ph.D.
Associate Professor,
School of Chemical Engineering,
Universiti Sains Malaysia,
Engineering Campus,
Pulau Pinang, Malaysia

Chapter 1

Gas Slippage in Tight Formations

Sherif Fakher and Abdelaziz Khlaifat

Abstract

In order to address the gas slippage for flow through tight formation, with a very low porosity (less than 10%) and permeability in micro-Darcy range, a series of single-phase gas flow experiments were conducted. Two different gases (N_2 and He) were used to carry out many single-phase experiments at different overburden and pressure drops and were compared with carbon dioxide (CO_2) flow types. The pore size distribution measurements showed the existence of a wide range of pore size distribution. Also, the single-phase gas flow experiments through the core plug, mostly at low pressure, showed Knudsen diffusion type, which is an indication of gas molecules' slippage at the wall of the pores.

Keywords: tight formation, slip effect, Knudsen diffusion, non-Darcy flow, pore size distribution

1. Introduction

Due to its high compressibility, gas flow behavior can vary greatly as the porous media size varies. Understanding this behavior and the type of flow is vital for the oil and gas industry, especially with the increase in production from unconventional reservoirs with extremely low permeability.

The flow behavior in pores can be estimated using many different mathematical models and experiments, the majority of which used the Knudsen number definition [1–3]. Many mathematical models were developed to determine the flow regime in nanopores as a function of adsorption [4, 5], rock permeability [6, 7], and molecular dynamics [8, 9]. All of these models used some form of the Knudsen number definition in their model in order to predict gas flow behavior as a function of different parameters. Many researchers also attempted to model Knudsen diffusion in shale gas reservoirs to determine the recovery potential when the dominant flow regime was Knudsen diffusion [10–13].

One of the main experimental methods to determine gas flow in nanopores relies on understanding the formation properties of the unconventional reservoir. Research has shown that unconventional tight sand gas reservoirs have three distinct features. These include relatively large pores with mineral deposition in the pores that resulted in a reduction in the overall pore diameter, narrow and flat pores that were generated due to alteration of the primary porosity of the rock, and grains supported by ultra-fine micro-matrix particles, usually clays [14–16]. All three of these have a common feature, which is an extremely small average pore diameter, reaching nanoscale in most cases.

This research studies the predominant flow regime that will be observed for gas flow in nanopores of low-permeability unconventional gas reservoirs using tight sandstone cores and three gasses including helium, nitrogen, and carbon dioxide. The research focuses on pores ranging from less than 10 nm in size up to 400 nm. The research relies on experimental analysis of the pores and the definition of the Knudsen number to determine the flow regime for all three gasses.

2. Knudsen flow regimes

To determine the flow behavior through pores, one of the most widely used methods is Knudsen number. It can be defined as a ratio between the mean free path and the pore diameter of the porous media through which the fluid is propagating. The mean free path is the distance that the gas molecule travels until it begins colliding with the other gas molecules. It can be defined mathematically as a function of fluid and thermodynamic conditions as shown in Eq. (1) [1–3]:

$$\lambda = \frac{3.2\mu}{p}\sqrt{\frac{PT}{2\pi M}} \tag{1}$$

where λ is the mean free path, μ is the viscosity, p is the pressure, T is the temperature, M is the fluid molecular weight, and R is the universal gas constant.

Knudsen number integrates the mean free path, which includes the fluid and the thermodynamic properties, with the pore size, which is a rock property, to predict the flow behavior of the gas in different pore sizes. Knudsen number is defined mathematically as shown in Eq. (2) [12]:

$$K_n = \frac{\lambda}{2r} \tag{2}$$

where K_n is the Knudsen number, and r is the average pore radius or half the pore diameter.

Based on the definition of Knudsen number, there are four possible situations for the gas flow in different pore sizes. These include the following:

- $K_n \geq 10$: This type of flow is referred to as Knudsen diffusion. This occurs when the mean free path is much larger than the pore diameter. This is the case in small pore sizes, less than 10 nm. Collison between both the gas molecules and the capillary walls is important in this case.

- $10 > K_n \geq 0.1$: This is known as a transition region. It highlights the beginning of the transition between Knudsen diffusion and normal flow. In this case, the large difference between the mean free path and the pore diameter begins narrowing down. The molecule-wall interactions begin reducing; however, its effect is still significant on the flow.

- $0.1 > K_n \geq 0.01$: This is known as slip flow. In some cases, this is grouped with the transition flow as one case due to the similarity in behavior. In this case, the mean free path and the pore diameter are comparable with each other. Knudsen number begins decreasing until it is very close to the range of viscous flow, although it has still not reached it. The molecule-molecule and molecule-wall interactions are still both significant in this case.

- $0.01 > K_n$: This case is referred to as viscous flow. In this case, the mean free path is much smaller than the pore diameter. The molecule-molecule collision becomes predominant, whereas the molecule-wall collision is insignificant. In this case, normal molecular diffusion occurs.

3. Experimental description

A description of the material, setup, and procedure followed to conduct all the experiments is explained in this section. All experiments were conducted using the same conditions to be able to accurately compare the results.

3.1 Experimental material

The material used to conduct the experiments include the following:

- *Distilled Water*: Distilled water was used for the displacement of fluids through the accumulator and for some of the saturation experiments.

- *Nitrogen Gas*: Nitrogen gas is provided through high-pressure nitrogen cylinders. The nitrogen was injected and pressurized in the high-pressure vessels for the experiments.

- *Helium Gas*: Helium gas is provided through high-pressure helium cylinders. The helium was injected and pressurized in the high-pressure vessels for the experiments.

- *Sandstone Core Plugs*: Sandstone core plugs from Travis Peak Formation in East Texas collected from a depth of 8700 ft were used to conduct the experiments. The cores were 3.5 inches in length and 1.5 inches in diameter.

- *Stainless Steel Tubing*: The stainless steel tubings were allocated in the setup to create connections between the different high-pressure vessels, the pump, and the core holder.

- *High-Pressure Valves*: High-pressure valves with a pressure limitation of 15,000 psi were used to control the flow of fluids across the setup.

- *Transducers*: High-accuracy transducers that could record pressure, temperature, and flow rate were used in the experiments to log these values for the duration of the experiment.

3.2 Experimental setup

The setup used to conduct the gas flow regime through low-permeability porous media is illustrated in **Figure 1**. The setup is composed of a pump used to inject the fluids into four different high-pressure vessels in the core holder. The vessels include water, helium, and nitrogen. The core holder is connected to transducers that measure pressure, temperature, and flow rates. The core holder houses a sandstone core used to mimic the porous media. Valves are allocated across the setup to control the flow direction of different fluids.

4. Results

The results for the pore size distribution using both mercury intrusion and gas sorption will be presented. Also, the dominant flow behavior for different pore size ranges will be determined for each of the three gases used in this research.

To determine the flow regime, the Knudsen number definition was used. Since it relies on the mean free path, this was calculated for all three gases. The mean free path for He, N_2, and CO_2 at ambient conditions is shown in **Table 1**.

4.1 Mercury intrusion

Mercury intrusion porosimetry technique involves the measurement of the capillary pressure and volume of mercury that penetrated the sample pores at a specified pressure. Although widely used to determine pore distribution, this method will not detect the pores that are present in necks narrower than the pore itself due to the inability of the mercury to intrude in them.

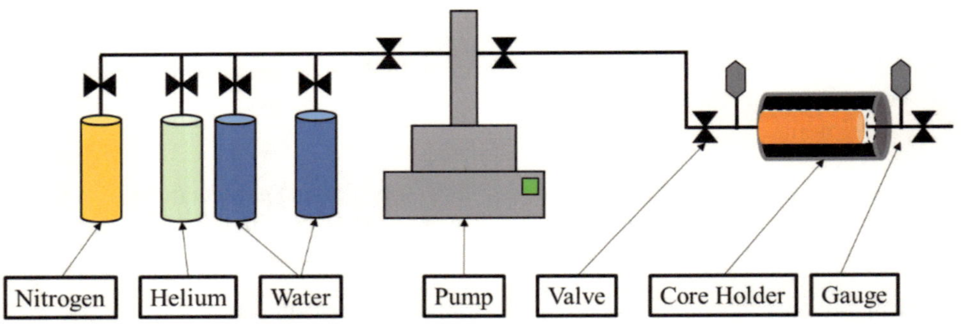

Figure 1.
Gas flow regime experimental setup.

Gas type	Mean free path, nm
Nitrogen	67
Helium	195.5
Carbon dioxide	28

Table 1.
Mean free path for N_2, He, and CO_2 at ambient conditions.

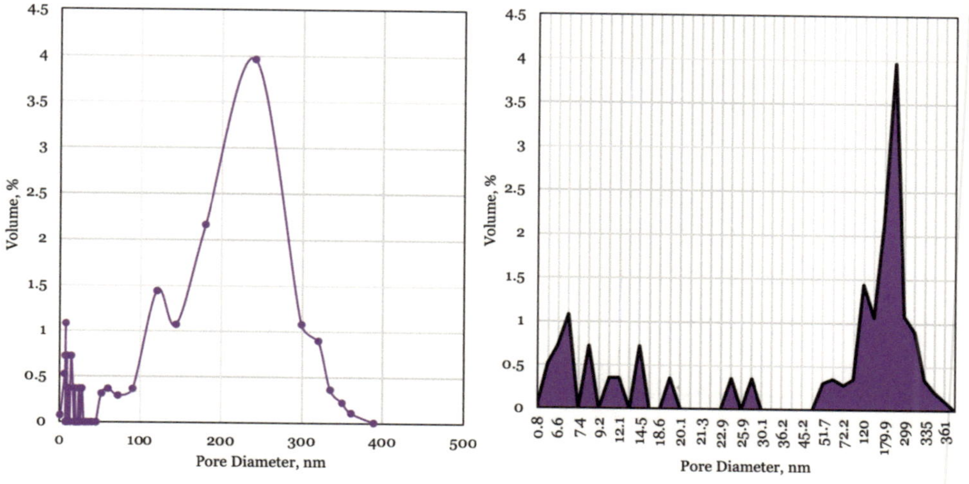

Figure 2.
Pore size distribution using mercury intrusion.

4.1.1 Pore size distribution

The pore size distribution using mercury intrusion is presented in **Figure 2**. A wide distribution of pores was available in the core, with the highest volume being for the 180 nm pore size. Based on the available pore sizes and their frequency, the Knudsen number was calculated to determine the predominant flow regime for the helium, nitrogen, and carbon dioxide.

4.1.2 Flow behavior in pores

The flow behavior of all three gasses in the core was evaluated using Knudsen number. Based on the ranges previously discussed, the flow behavior was determined for different pore sizes. The Knudsen number was plotted versus the pore size for three different ranges of pore sizes including up to 10 nm, between 10 and 100 nm, and greater than 100 nm.

The Knudsen number values for the pore size up to 10 nm for all three gasses are shown in **Figure 3**. For all pore sizes, helium exhibits Knudsen diffusion. This is predominantly due to the helium molecular size being small compared with the nitrogen and the carbon dioxide. As for the nitrogen and the carbon dioxide, their flow behavior is very similar, as both exhibit a transition from low Knudsen number to large Knudsen number for the same pore diameter. The main predominant flow behavior for both gasses is transition flow and Knudsen diffusion. This is due to the larger molecular size of both compared with helium.

The Knudsen number values for the pore sizes between 10 and 100 nm for all three gasses are shown in **Figure 4**. It can be observed that the Knudsen number values for all three gasses are decreasing with the increase in pore size. This is due to the increase in the pore diameter relative to the mean free path. For pore sizes greater than 20 nm, Knudsen diffusion disappears entirely for all three gasses. For helium, the flow is dominated by transition flow, whereas for the nitrogen and carbon dioxide, the flow is dominated by transition and slip flow.

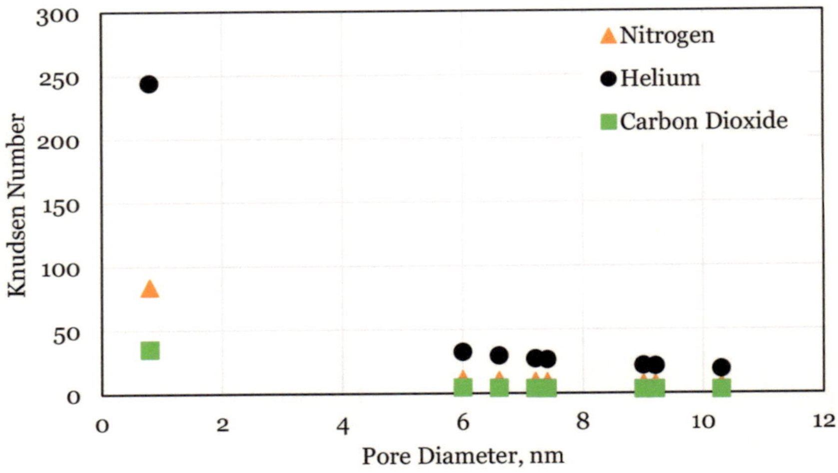

Figure 3.
Knudsen number for all three gasses for average pore sizes less than 10 nm using mercury intrusion method.

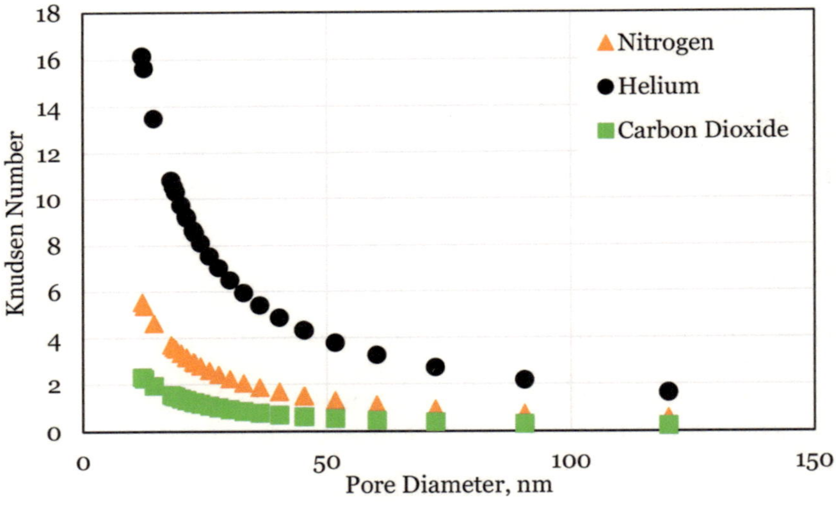

Figure 4.
Knudsen number for all three gasses for average pore sizes between 10 and 100 nm using mercury intrusion method.

The Knudsen number values for the pore sizes greater than 100 nm for all three gasses are shown in **Figure 5**. For helium, this pore size marks the transition from transition flow to slip flow. As for the nitrogen and carbon dioxide, this pore size range marks the dominance of slip flow and the appearance of molecular diffusion. Since the pore diameter is large compared with the mean free path, the molecule-wall interaction is small compared with the molecule-molecule interaction.

4.2 Gas sorption isotherm

A cylindrical pore model was used to determine the pore size distribution using gas sorption isotherm. This method is accurate for pore sizes ranging between

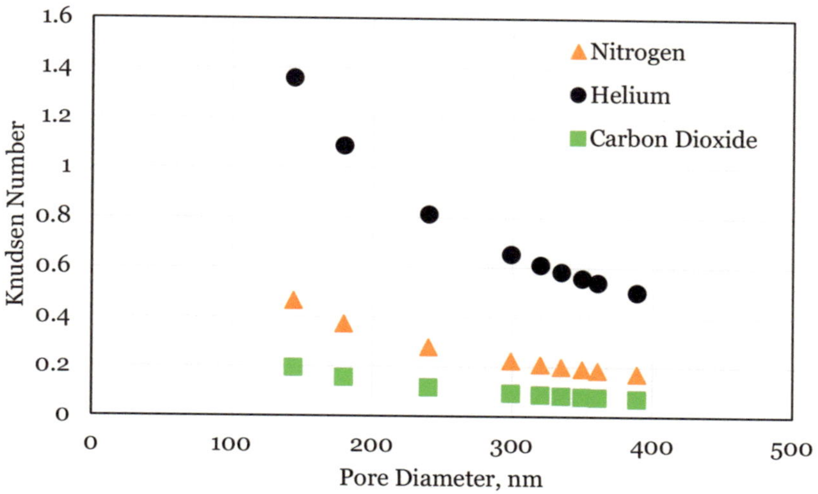

Figure 5.
Knudsen number for all three gasses for average pore sizes greater than 100 nm using mercury intrusion method.

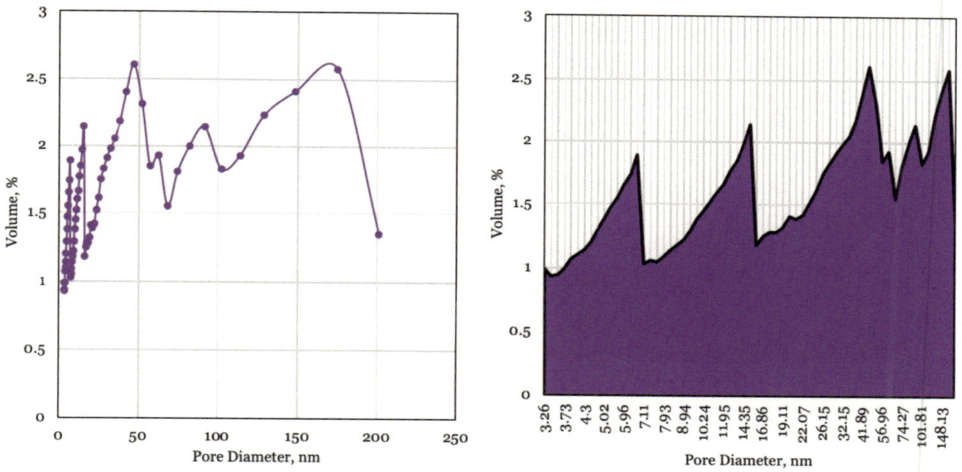

Figure 6.
Pore size distribution using gas sorption.

0.4 and 200 nm. The pore size distribution was determined by analysis of the adsorption isotherm using Kelvin's equation to correlate the adsorbing gas relative pressure in equilibrium with the porous media.

4.2.1 Pore size distribution

The pore size distribution for the core sample used in this research using gas sorption is shown in **Figure 6**. Compared with the pore size distribution using mercury intrusion, the largest pore size found using the gas sorption method was in the range of 200 nm. The maximum volume of pore sizes was 56 nm.

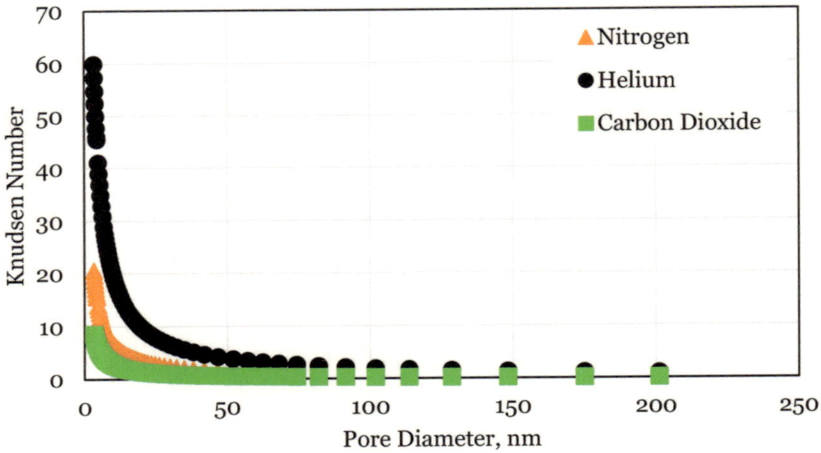

Figure 7.
Knudsen number for all three gasses using gas sorption.

4.2.2 Flow behavior in pores

Similar to the flow behavior analysis using the pore size distribution found via mercury intrusion, the flow regime for all three gases was evaluated using Knudsen number for the pore size distribution obtained using gas sorption. **Figure 7** shows the overall Knudsen number values for all the pore sizes for all three gases. These are then divided into three pore size distributions for clearer analysis. The ranges include pore sizes less than 10 nm, pore sizes between 10 and 100 nm, and pore sizes greater than 100 nm.

The Knudsen number values for all three gasses in pores diameters less than 10 nm are shown in **Figure 8**. For helium, the flow is entirely Knudsen diffusion for diameters

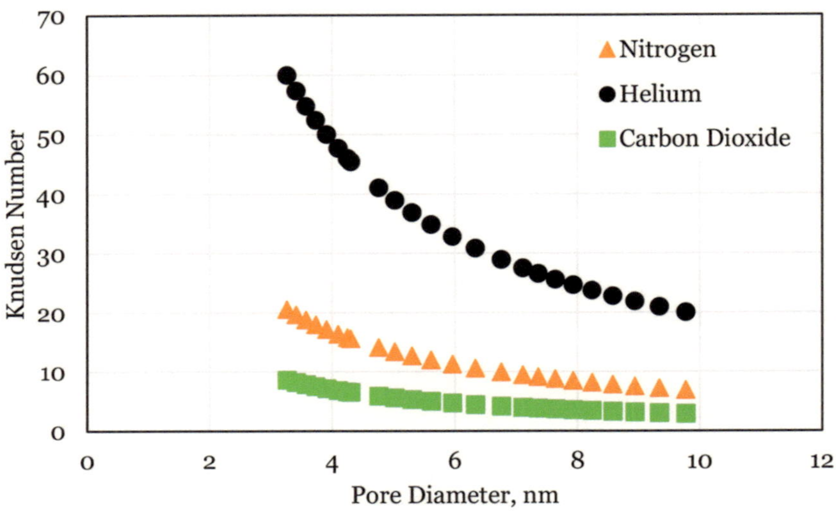

Figure 8.
Knudsen number for all three gasses for average pore sizes less than 10 nm using gas sorption.

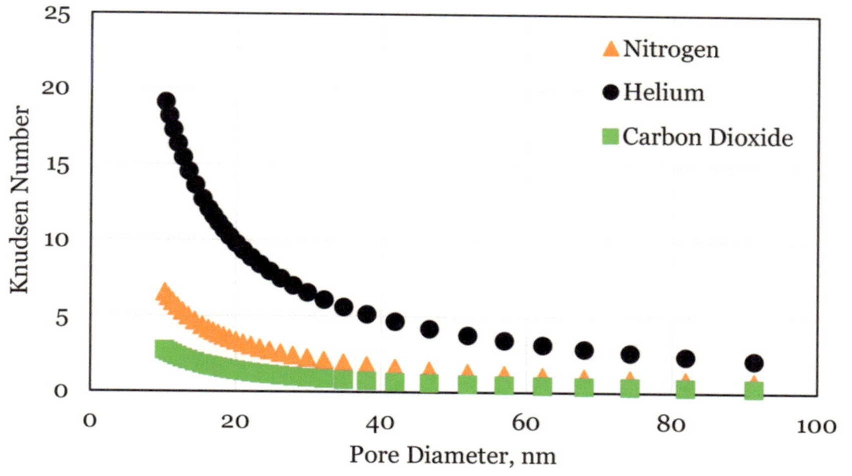

Figure 9.
Knudsen number for all three gasses for average pore sizes between 10 and 100 nm using gas sorption.

up to 10 nm. Nitrogen exhibits Knudsen diffusion until the pore diameter reaches 6 nm, above which the flow begins to be transition flow. As for the carbon dioxide, the flow begins at transition and does not reach Knudsen diffusion. Since the pore radius is uniform for all three gasses, the main variation that results in the change in the flow regime is the mean free path. The thermodynamic conditions are also the same for all experiments; therefore, the main factor contributing to the variation in the mean free path value is the size of the gas molecule.

The Knudsen number values for all three gasses in pores diameters between 10 and 100 nm are shown in **Figure 9**. The flow behavior for helium begins to shift to transition flow when the pore diameter exceeds 20 nm. For nitrogen and carbon dioxide, the flow behavior is dominated by transition flow. As the pore sizes increase beyond 80 nm, the flow behavior begins approaching slip flow.

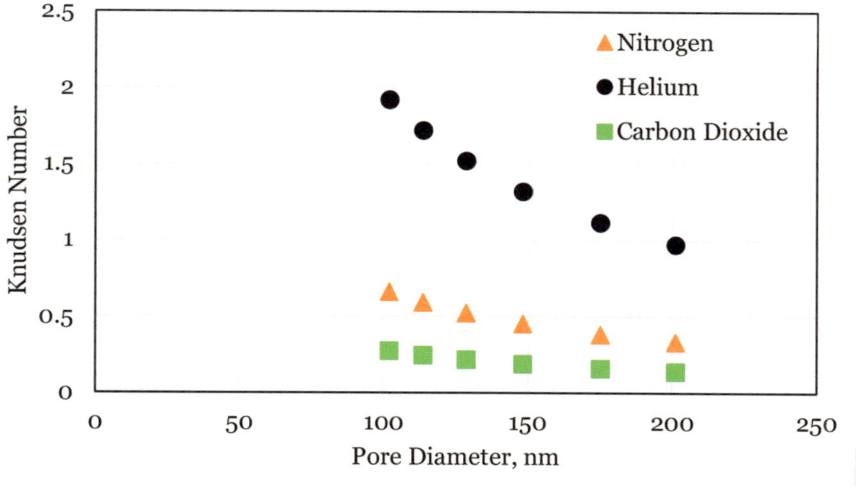

Figure 10.
Knudsen number for all three gasses for average pore sizes greater than 100 nm using gas sorption.

Measurement method	Average pore size range, nm	Gas type	Dominant flow type
Mercury Intrusion	Less than 10	He	Knudsen diffusion
	10–100		Transition flow
	Greater than 100		Transition flow Slip flow
	Less than 10	N_2	Knudsen diffusion Transition flow
	10–100		Transition flow Slip flow
	Greater than 100		Slip flow
	Less than 10	CO_2	Knudsen diffusion Transition flow
	10–100		Transition flow Slip flow
	Greater than 100		Slip flow Molecular diffusion
Gas Sorption	Less than 10	He	Knudsen diffusion
	10–100		Knudsen diffusion Transition flow
	Greater than 100		Transition flow
	Less than 10	N_2	Knudsen diffusion Transition flow
	10–100		Transition flow
	Greater than 100		Transition flow
	Less than 10	CO_2	Transition flow
	10–100		Transition flow
	Greater than 100		Transition flow Slip flow

Table 2.
Knudsen number values for all experiments.

The Knudsen number values for all three gasses in pores diameters greater than 100 nm are shown in **Figure 10**. For all three gasses, transition flow is the dominant flow behavior in this pore size. The flow regime for carbon dioxide reaches slip flow in the largest pore size observed, whereas, for all other pore sizes, the flow is dominated by transition flow.

4.3 Summary of flow behavior for all three gasses

Based on the analysis of the flow regimes for helium, nitrogen, and carbon dioxide using pore size distribution obtained using mercury intrusion and gas sorption, the predominant flow regime for each gas was determined. In order to determine the change in flow regime with pore size, the pore sizes were analyzed using three different ranges. A summary of the dominant flow regimes for each gas type in different pore sizes is shown in **Table 2**.

5. Conclusion

This research investigates the dominant flow regime for three gasses, including helium, nitrogen, and carbon dioxide, in nanopore sizes to determine the most prominent flow behavior present in unconventional gas reservoirs. The research utilizes the Knudsen number definition to determine the flow behavior as a function of the pore diameter size and the mean free path. Based on the results, it was found that in pore sizes less than 10 nm, the dominant flow behavior was Knudsen diffusion, while in pore sizes greater than that, the flow behavior begins to transition from diffusion dominated to viscous-dominated flow, until eventually in the larger pore sizes, the flow is dominated by molecular diffusion.

Author details

Sherif Fakher and Abdelaziz Khlaifat*
The American University in Cairo, Cairo, Egypt

*Address all correspondence to: abdelaziz.khlaifat@aucegypt.edu

References

[1] Khlaifat A. Non-Darcy flow in tight gas reservoir of travis peak formation. In: Paper presented at the SPE Middle East Unconventional Gas Conference and Exhibition. Muscat, Oman; 2011

[2] Khlaifat A, Qutob H, Arastoopour H. Deviation from Darcy's flow in fractured tight gas sand reservoirs. In: Paper presented at the SPE/DGS Saudi Arabia Section Technical Symposium and Exhibition. Al-Khobar, Saudi Arabia; 2011

[3] Al-Khlaifat A. Two-phase flow through low permeability fractured tight sand porous media (Order No. 9833028). Available from ProQuest Dissertations & Theses Global. (304439510). 1998

[4] Shabro V et al. Numerical simulation of shale-gas production: From pore-scale modeling of slip-flow, Knudsen diffusion, and langmuir desorption to reservoir modeling of compressible fluid. In: Presented at the SPE North American Unconventional Gas Conference and Exhibition. Woodlands, Texas, USA; 2011

[5] Shabro V et al. Forecasting gas production in organic shale with the combined numerical simulation of gas diffusion in Kerogen, Langmuir desorption form kerogen surfaces, and advection in nanopores. In: Presented at the 2012 SPE Annual Technical Conference and Exhibition. San Antonio, Texas, USA; 2012

[6] Gao C et al. The shale-gas permeability measurement considering the rarefaction effect on transport mechanism in the nanopores. In: Presented at the International Petroleum Technology Conference. Beijing, China; 2013

[7] Niu C et al. Second-order gas-permeability correlation of shale during slip flow. SPE Journal. 2014;**19**:786-792. DOI: 10.2118/168226-PA

[8] Okamoto N et al. Slip velocity and permeability of gas flow in nanopores for shale gas development. In: Presented at the SPE Asia Pacific Unconventional Resources Conference and Exhibition. Brisbane, Australia; 2015

[9] Okamoto N et al. Slip velocity of methane flow in nanopores with Kerogen and quartz surfaces. SPE Journal. 2018;**23**: 102-116. DOI: 10.2118/176989-PA

[10] Wu K et al. A unified model for gas transfer in nanopores of shale-gas reservoirs: Coupling pore diffusion and surface diffusion. SPE Journal. 2016;**21**:1583-1611. DOI: 10.2118/2014-1921039-PA

[11] Xu J et al. Nanoscale free gas transport in shale rocks: A hard-sphere based model. In: Presented at the SPE Unconventional Resources Conference. Alberta, Canada; 2017

[12] Fakher S et al. What are the dominant flow regimes during carbon dioxide propagation in shale reservoirs' matrix, natural fractures and hydraulic fractures? In: SPE Western Regional Meeting, Virtual. 2021

[13] Fakher S et al. Increasing oil recovery from unconventional shale reservoirs using cyclic carbon dioxide injection. In: Paper presented at the SPE Europec, Virtual. 2020

[14] Sharer JC, O'Shea P. GRI's research program on unconventional natural gas. Chemical Engineering Program. 1986;**82**:19

[15] Seoder DJ. Assessing tight gas reservoir quality using petrography and core analysis. In: Proceeding of International Gas Research Conference. 1989

[16] Seoder DJ. Reservoir properties and the pore structure of tight gas sands. AAPG Bulletin. 1984;**64**:4

Chapter 2

Rheology of Heavy Oils

Gudret Isfandiyar Kelbaliyev, Dilgam Babir Tagiyev and Manaf Rizvan Manafov

Abstract

The problems of heavy oil rheology, accompanied by physical phenomena of the formation and destruction of coagulation disordered structures and aggregates as a result of the hydrodynamic interaction of particles (asphaltenes, paraffins, resins, and solid-phase particles) contained in the oil, which significantly affect its properties and flow, are considered and analyzed. Rheological models of viscous-plastic heavy oils are considered and developed, consistent with a variety of experimental data. New rheological models for viscous-plastic heavy oils are proposed, which make it possible to generalize many existing models. It is noted that the variety of rheological models for heavy oils is determined by the conditions for the formation of disordered structures in the bulk of the oil flow. For heavy oils, a nonlinear equation for filtration in porous media is proposed, depending on the shear stress, pressure gradient, effective viscosity of the oil, and a number of other parameters. An analytical solution to this equation is proposed, which is consistent with the experimental data. Models for the settling rate and drag coefficient of particles in heavy oils are proposed. Applied problems of rheology aimed at improving the rheological properties of heavy oil during their processing as a result of creating a recirculation scheme at an operating oil refining unit are considered.

Keywords: rheology, heavy oils, effective viscosity, filtration, structure formation, asphaltenes, rheological models, production, processing, settling, drag coefficient, recycling

1. Introduction

Crude oil, due to the content of various particles of different nature and properties, shapes, and sizes, is an oil dispersed system, with their inherent physical and chemical phenomena of physical interaction between particles, structure formation, particle settling, and stratification of the entire system, affecting the phenomena of all types of substance transfer [1–3]. The rheology of oil and oil products is a field of science that studies and quantitatively predicts the formation and transformation of the state of oil dispersed systems over time. Heavy oils with a high content of various impurities (water, solid-phase), asphalt-resin compounds, and various paraffin are prominent representatives of media with rheological properties.

In the practice of oil production, transportation, refining, and the use of petroleum products, it is necessary to solve various, sometimes opposite problems of regulating the structural and mechanical properties of oil dispersed systems. The rheological properties of oil dispersed systems are defined as a high content of dispersed particles (water drops, gas bubbles, and solid particles), as well as asphalt-resin and paraffin particles dissolved in heavy oil, and their ability to form various structures as a result of the physical interaction of particles with each other. An important role in structure formation and the construction of rheological models of non-Newtonian fluids is played by such factors as the size and shape of particles, their concentration, and properties. In works [1–6], a lot of rheological models are given that describe the flow of oils from various fields, characterized by different properties. As is known, the nature of the relationship between shear stress τ, shear strain $\dot{\gamma}$ and their changes in time $d\tau/dt$ and $d\dot{\gamma}/dt$ is the essence of rheological research [2–5]. Rheological parameters are used as a criterion in the classification of high-viscosity viscous-plastic oils and heavy products of its processing. The mechanism of plastic flow consists of a set of acts of rupture and restoration of contacts in aggregates and coagulation structures, between dispersed particles after overcoming the limiting shear stress τ_0 of the system under study. A special class is the problems associated with the rheology of non-Newtonian oils or oil dispersed systems containing, in addition to particles of the solid phase, liquid drops of water and gas bubbles, and asphaltenes, resins, and at low temperatures paraffin particles dissolved in oil. The presence of such a spectrum of particles of various sorts and nature leads to the formation of very complex coagulation structures, which later turn into aggregates, clusters of aggregates, up to the formation of a viscoelastic framework, at which the flow velocity is zeroed. In this case, the viscosity of the oil system increases sharply with an increase in the content of particles, and accordingly the shear stress increases to a limiting value. At values of shear stress greater than the elastic limit stress $\tau > > \tau_0$, the structure is destroyed and the viscosity drops sharply. Rheological models of oil dispersed systems allow one to judge the fundamental properties of the systems themselves. These models, due to the lack of a common mechanism of flow and deformation, are mainly of an empirical or semiempirical nature. Despite a large number of empirical rheological models of structured disperse systems, there are currently no qualitative and quantitative theories linking the rheological properties of the system with the structure parameters. It is possible that in each particular case this is due to the nature of the flow, deformation, and formation of a disordered structure. Therefore, the results show that the same experimental study can be described by different empirical or semiempirical equations with the same model accuracy.

Heavy oils with a high content of asphalt-resinous substances are viscous-plastic liquids and are mainly described by the following rheological models [4–8]: a) Bingham model $\tau = \tau_0 + \eta\dot{\gamma}$, b) Ostwald-de Ville model $\tau = k_0\dot{\gamma}^n = k_0\dot{\gamma}^{n-1}\dot{\gamma}$, c) Hershel-Bulkley model $\tau = \tau_0 + k_0\dot{\gamma}^n$, and other modified type equations, where k_0 is the consistency coefficient and $\dot{\gamma}$is the shear rate.

The purpose of this work is to build a rheological model of heavy oils, accompanied by structure formation by particles contained in the volume, free settling of particles, and oil filtration through anisotropic porous media.

2. Structure formation in oil dispersed media

Coagulation structures are formed due to intermolecular bonds between particles, and if liquid interlayers remain between the particles, then the thickness of this interlayer significantly affects the strength of the coagulation structure. Aggregative unstable oil systems are characterized by the variability of the state of the environment, due to continuous structure formation and changes in the physical properties of particles, that is, a change in the volume and size of asphaltene particles as a result of their interaction, collision, coagulation, and crushing at a certain concentration in a closed volume. The relationship between the structure and viscosity of petroleum dispersed systems, as well as the features of their non-Newtonian flow, are explained by a change in the structure as a result of the formation and destruction of aggregates from asphaltene particles in the presence of resins. Oil structured systems containing crystals of high molecular weight paraffin, resins, and asphaltene particles, and at very low laminar flow rates or in the absence of flow form a chain or, in the limiting case, a continuous grid (framework). Sequential coagulation or agglomeration of individual asphaltene nanoparticles into nanoaggregates and clusters of nanoaggregates eventually leads to the formation of a viscoelastic framework that imparts certain rheological properties to heavy oils. The paper notes that real oil dispersed systems are classified according to activation energies into two structural groups that differ in the nature of the intermolecular interaction of particles in an oil disperse medium. These groups differ from each other in the content of asphaltenes and resins, and they can be classified into immobile with a low content of asphaltenes and interacting with a high content of asphaltenes. **Figure 1** shows the characteristic changes in activation energies for the two indicated groups.

The rheological model of the flow of oil dispersed systems can be based on the following assumptions [2, 7]:

- In structured oil system, there are nanoaggregates that have arisen as a result of collision, coagulation, and aggregation of asphaltene particles due to diffusion in laminar and turbulent shear flow and sedimentation (gravitational coagulation), the formed asphaltene aggregates can be deposited on the surface, forming a rather thick layer of deposits on the walls of the porous oil reservoir. Moreover, pressure drops depending on temperature can lead to redissolution or detachment of particles of precipitated asphaltenes during intensive mixing or turbulent flow;

- Nanoaggregates move as independent flow units until they collide with other similar aggregates or asphaltene particles;

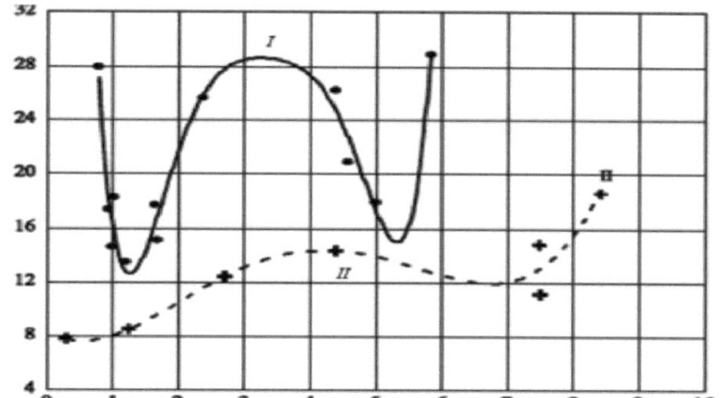

Figure 1.
The dependence of the activation energy on the content of asphaltenes for the immobile group (I) and interacting group (II) [2].

- Nanoaggregates, when colliding with each other, unite into clusters of nanoaggregates and then create a viscoelastic frame of a disordered structure with the highest possible viscosity and a loose coagulation structure. The maximum size of the framework of nanoaggregates is determined by the dimensions of the channels (pores and pipes) through which the flow flows. It is important to note that the formation of disordered structures in the volume of oil is the reason for the diversity of rheological models;

- Nanoaggregates can rotate in a gradient field and break under the action of tensile hydrodynamic forces depending on the pressure gradient or flow rate;

- The linear dimensions of nanoaggregates are in the range of the size of an individual asphaltene particle up to the maximum size of a cluster or framework of a disordered structure;

- In the limiting case of infinite velocity, all aggregates, under the condition $\lim_{\tau\to\infty}(\tau_0/\tau) \to 0$, are destroyed to individual particles as a result of which the flow of a dispersed system approaches Newtonian;

- In the presence of aromatic hydrocarbons, asphaltenes dissolve well, thereby preventing structure formation , i.e. formation of clusters and viscoelastic framework. The solubility of asphaltenes is affected by the presence of other compounds contained in the oil, such as resins.

The rheological equation of Maxwell's viscoelastic fluid in substantial derivatives is written as:

$$\lambda\left(\frac{\partial\tau}{\partial t} + U\frac{\partial\tau}{\partial y}\right) + \tau = \eta_c\dot{\gamma} \tag{1}$$

$$t = 0,\ \tau = \tau_0,\ \dot{\gamma} = 0$$

where λ is the relaxation time.

The Eq. (1), presented in the form:

$$\lambda\left(\frac{\partial\tau}{\partial t} + U\frac{\partial\tau}{\partial y}\right) + \tau = 0 \tag{2}$$

The solution of Eq. (2) can be represented as:

$$\tau = C_1 f(y - Ut)\exp\left(-{}^{\tau}\!/\!_{\lambda}\right) \tag{3}$$

Substituting this solution into Eq. (2), we obtain an identity. Here, $\lambda = \eta_c/G-$ is the Maxwell relaxation time, U is the strain front movement velocity, $f(y - Ut)$ is a function that determines the strain front in the frame, y is the coordinate, G *is* shear modulus, $\dot{\gamma} = d\gamma/dt-$ is shear rate, $\gamma-$ is the shear gradient, τ_0 is the ultimate shear stress or yield strength. Moreover, if $\tau \le \tau_0$, then $\dot{\gamma} = 0$. The complete solution of Eq. (3) takes the form:

$$\tau = C_1 f(y - Ut)\exp\left(-{}^{t}\!/\!_{\lambda}\right)\tau_0 \tag{4}$$

or this equation can be written in logarithmic form:

$$\ln \tau = \ln \tau_0 - t/\lambda + \ln (C_1 f(y - Ut)), \quad \tau_0 = \eta_c \dot{\gamma} \tag{5}$$

It is obvious that the value t/λ in Eq. (5) characterizes the deformation of the viscoelastic frame in time and, with the flow of heavy oils, depends on the velocity or pressure gradient, which in the approximation can be represented as: $t/\lambda = t\dot{\gamma}/\mathrm{Wi} \sim f\left[(\mathrm{grad}P/(\mathrm{grad}P)_0)^n\right]$

(where $\mathrm{Wi} = \lambda\dot{\gamma}$ is the Weissenberg number).

3. Rheology of viscoplastic heavy oils

Heavy oils with constant differential viscosity are characterized by viscous-plastic properties. Bingham viscous-plastic fluids include petroleum dispersed systems, polymeric fluids, many types of food materials, cement mortars, oil paints, and others that exhibit viscous, plastic, and highly elastic properties. They differ from conventional liquids in that some finite stress is required to initiate flow. The rheology of viscoplastic fluids is described by the Bingham equation.

$$\tau = \tau_0 + \eta\dot{\gamma}, \quad \dot{\gamma} > 0 \tag{6}$$

Moreover, if $\dot{\gamma} = 0$, then $\tau \leq \tau_0$. In this equation, the shear rate can be represented as the flow velocity gradient:

$$\dot{\gamma} = \frac{d(dx/dy)}{dt} = \frac{d(dx/dt)}{dy} = \frac{dV_x}{dy} \tag{7}$$

Based on expression (6), the viscosity of viscous-plastic oil can be determined as:

$$\eta = \frac{\tau - \tau_0}{\dot{\gamma}} \tag{8}$$

However, some viscous-plastic fluids exhibit properties that do not obey Eq. (6). Such fluids are usually described by other nonlinear rheological equations of the Ostwald-de Ville or Herschel-Bulkley type, which are widely used to describe the flow of plastic fluids, heavy petroleum oils, and petroleum dispersions. This is explained by the presence of various particles of the dispersed phase in the liquid, and therefore, such systems are prone to the formation of coagulation structures up to the formation of a cluster of aggregates and a framework under the condition $\tau < \tau_0$.

In addition, the coefficients included in the rheological equation will depend on the concentration, size, and properties of particles, as well as on temperature and many other parameters.

Currently, there is no consensus on the mechanism of non-Newtonian flow of oil dispersed systems, and therefore, the set of flow equations $\tau(\gamma^\bullet)$ or rheological viscosity equations $\eta(\gamma^\bullet)$ *or* $\eta(\tau)$ used in practice are mainly empirical or semiempirical. However, despite a large number of works and a variety of approaches in the field of rheology of structured disperse systems, including oil disperse systems, there is still no satisfactory quantitative theory linking the rheological properties of bodies with their structural parameters. It is possible that this is due to the formation of various disordered structures that affect the type of rheological model and the nature of the dependence of the effective viscosity of the disperse system on stress and shear rate.

Let us assume that the change in relative viscosity is proportional to linear deformation in a nonlinear form:

$$\frac{\Delta\eta}{\eta} \sim m_0 \left(\frac{\Delta x}{\Delta y}\right)^p \tag{9}$$

where Δx is the increment of linear deformation with a change in Δy and m_0 is the coefficient of proportionality.

In the limiting case $\Delta y \to 0$, passing to the differential form, we take the nonlinear form of the expression (9)

$$\frac{d\eta}{\eta} = m_0 \left(\frac{dx}{dy}\right)^{p-1} d\left(\frac{dx}{dy}\right) \tag{10}$$

Here p is the exponent characterizing the degree of nonlinearity. Assuming that $dx/dy = \lambda dV_x/dy = \lambda\dot{\gamma}$, we have:

$$\frac{d\eta}{\eta} = m_0 (\lambda\dot{\gamma})^{p-1} d(\lambda\dot{\gamma}) \tag{11}$$

Taking into account the initial and infinite viscosity, the solution, Eq. (11) can be written as:

$$\eta - C = (\eta_0 - C) \exp\left(\frac{m_0}{p}(\lambda\dot{\gamma})^p\right) \tag{12}$$

where λ is relaxation time. Moreover, if $\dot{\gamma} \to \infty$, then $C = \eta_\infty$. As a result, the dependence of the viscosity of fluid on the shear stress in the general case for $m_0 = 1 - n$, we obtain the dependences of the viscosity on the shear rate for a dilatant $(n < 1)$, viscous-plastic fluid $(n > 1)$, and a Newtonian fluid $(n = 1)$ in the form.

$$\begin{gathered}
\frac{\eta - \eta_\infty}{\eta_0 - \eta_\infty} = \exp\left(\frac{m_0}{p}(\lambda\dot{\gamma})^p\right), \quad n > 1, \ m_0 > 0 \\
\frac{\eta - \eta_\infty}{\eta_0 - \eta_\infty} = \exp\left(-\frac{m_0}{p}(\lambda\dot{\gamma})^p\right), \quad n < 1, \ m_0 < 0 \\
\eta = \eta_0, \quad n = 1, \ m = 0
\end{gathered} \tag{13}$$

We represent the last expression in the form:

$$\frac{\eta - \eta_\infty}{\eta_0 - \eta_\infty} = \exp\left(-\frac{m_0}{p}(\lambda\dot{\gamma})^p\right) = \frac{1}{\left[\exp\left((\lambda\dot{\gamma})^p\right)\right]^{m_0/p}} \tag{14}$$

Expanding the exponent in a series $\exp\left((\lambda\dot{\gamma})^p\right) = 1 + (\lambda\dot{\gamma})^p + \frac{1}{2}(\lambda\dot{\gamma})^{2p} + \ldots \ldots$, we get

$$\frac{\eta - \eta_\infty}{\eta_0 - \eta_\infty} = \frac{1}{\left[1 + (\lambda\dot{\gamma})^p + \frac{1}{2}(\lambda\dot{\gamma})^{2p}\right]^{m_0/p}} \tag{15}$$

Most of the formulas for determining the viscosity of a viscous-plastic fluid, taking into account the linear expansion of the exponential, are based on Ref. [9]. In particular, for describing the rheology of viscous-plastic polymeric fluids, the most effective is the Carreau-Yasuda rheological model [10, 11] presented in the form:

$$\frac{\eta - \eta_\infty}{\eta_0 - \eta_\infty} = \left[1 + (\lambda\dot{\gamma})^p\right]^{(n-1)/p} \tag{16}$$

Here, η_0, η_∞ are the value of the fluid viscosity at the beginning and at infinity, λ is the relaxation time, and p is a dimensionless coefficient characterizing the transition from the region with initial viscosity to the region with final viscosity. As noted in Ref. [10], Eq. (16) describes the rheology of polymeric liquids at various concentrations and temperatures. In Ref. [10], a simpler form is also considered for describing viscous-plastic fluids at $\eta_\infty = 0$

$$\eta = \frac{\eta_0}{1 + \lambda\dot{\gamma}} \tag{17}$$

In Ref. [12] for viscous-plastic fluids, Bingham or Herschel–Bulkley proposed the following rheological models.

$$\frac{\eta - \eta_\infty}{\eta_0 - \eta_\infty} = \frac{1}{1 + \left(\frac{|\dot{\gamma}|}{\dot{\gamma}^m}\right)^m} \tag{18}$$

$$\eta = \begin{Bmatrix} \tau_0 + k_0|\dot{\gamma}|^m, & \tau > \tau_0 \\ \infty & \tau \leq \tau_0 \end{Bmatrix} \tag{19}$$

It is also important to note the dependence of viscosity on shear stress in the following empirical formula (13)

$$(\eta - \eta_\infty)/(\eta_0 - \eta_\infty) = \exp\left(-(\lambda\dot{\gamma})^n\right) \tag{20}$$

Figure 2 shows the dependence of viscosity on the shear rate for some viscous-plastic fluid.

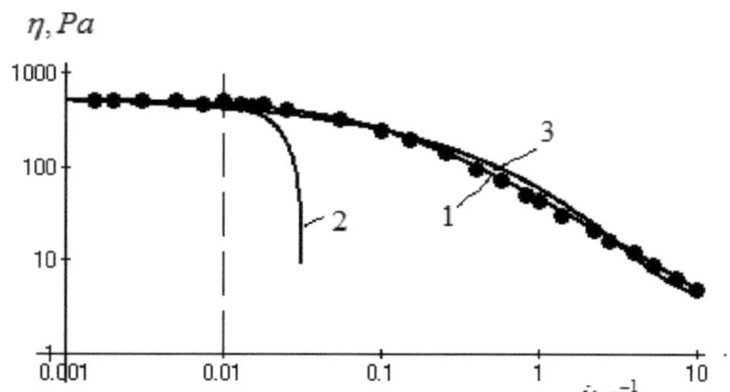

Figure 2.
Dependence of fluid viscosity on shear rate for various equations: 1-(17), $\eta_o = 500\,Pas$, $\lambda = 10s$; 2–$\eta = \eta_o - b\dot{\gamma}^2$ [1]; 3–$(\eta - \eta_\infty)/(\eta_o - \eta_\infty) = \exp\left(-2.3\dot{\gamma}^{0.45}\right)$.

In [13], the viscosity of a non-Newtonian polymer fluid containing particles of various concentrations is determined depending on the shear stress in the form (**Figure 3**).

$$\eta = \frac{\eta_0}{1 + (m\,\tau)^n} \tag{21}$$

It should be noted that the coefficients η_0, m, n included in Eq. (12) depend on the concentration (volume fraction), particle material, and temperature.

Figure 4 shows the curves describing the experimental values [12] of changes in the viscosity of polymeric liquids by Eqs. (15) and (16).

The above calculations and comparison with experimental data allow us to confirm the correctness of the accepted hypothesis about the proportionality of the relative viscosity of deformation (9).

In principle, rheological models for various flows of a non-Newtonian fluid do not obey physical laws but are empirical and semiempirical approximations and formulas that describe flow curves in a certain range of shear rates. Experimental

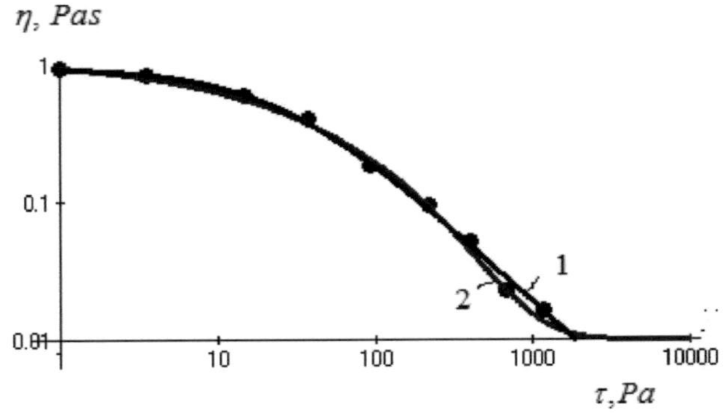

Figure 3.
Dependence of viscosity on shear stress for various equations: 1– $\eta = 0.95/(1 + (0.046\tau)^{1.055})$*:* 2 $-\eta = 1.15\exp(-0.21\tau^{0.47}) + 0.01$.

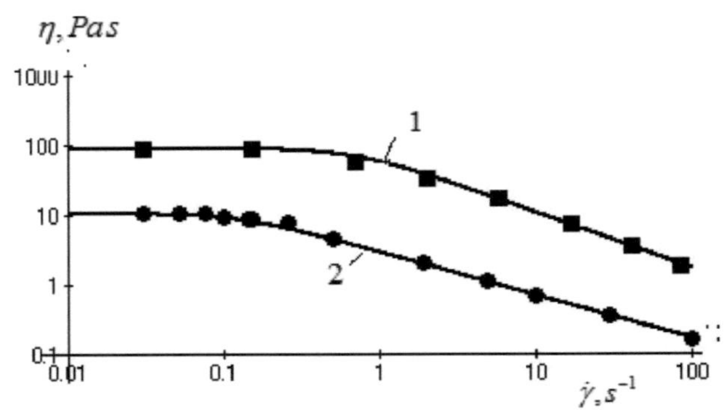

Figure 4.
Dependence of viscosity of polymeric fluids on shear rate: 1–0.75% polyacrylamide in 95/5 mixture by weight of water and glycerin ($p = 2, m_o/p = 0.4$ *); 2–7% aluminum soap in decalin and m-cresol (* $p = 2, m_o/p = 0.3$ *).*

measurements are approximated by certain approximate empirical or semiempirical equations, and the choice of the most convenient of them is largely determined by the degree of maximum approximation of calculated and experimental measurements or by the simplicity of the formulas used when solving applied problems.

4. Dependence of the effective viscosity of oil on the content of asphaltenes

An experimental study of the influence of the content of asphaltenes and resins in oil on its rheological properties and viscosity was proposed in the works [13–15].

Using the results of these studies, it can be noted that the presence of asphaltenes, resins, and paraffins in oil, which change the properties of the oil, significantly affects their movement and transport. First of all, this is reflected in the stress and shear rate and the increase in the viscosity of non-Newtonian oil. **Figure 5** suggests the dependence of the effective viscosity of Iranian oil on the shear rate by various rheological models [16].

The **Table 1** shows the values of the main coefficients included in these rheological models at various temperatures [16].

Of all the models, a satisfactory approximation to the experimental data is given by the expression.

$$\eta = 45.86\dot{\gamma}^{0.75} \tag{22}$$

Given this expression, the rheological dependence that satisfies the experimental data can be represented as:

$$\tau = \tau_0 + 45.76\dot{\gamma}^{1.75} \tag{23}$$

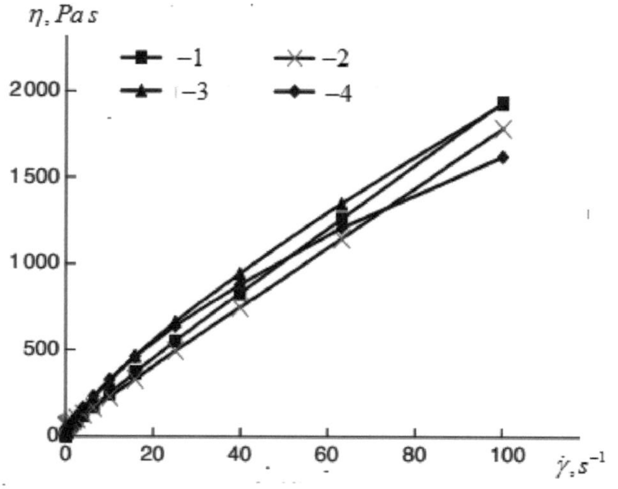

Figure 5.
Approximation of the dependence of viscosity on shear rate by various rheological models: 1 - Casson model – $\tau = \left(\tau_0^{1/2} + |k^{1/2}\dot{\gamma}^{1/2}\right)^2$; *2 -Bingham model –* $\tau = \tau_0 + \eta\dot{\gamma}$; *3 - exponential function –*$\tau = k\dot{\gamma}^n$; *4 - experiment.*

	Temperature								
	25⁰C			45⁰C			60⁰C		
Model	τ_0, Pa	k, Pas	n	τ_0, Pa	k, Pas	n	τ_p, Pa	k, Pas	n
Casson	8.13	4.11	–	0.57	2.2	–	0.15	1.53	–
Power law	–	54.65	0.77	–	9.26	0.88	–	3.86	0.9
Bingham	61	17.23	–	8.66	4.81	–	3.13	2.33	–

Table 1.
Coefficients of rheological models at temperatures $T = 25°C, 45°C$ and 60°C.

The dependence of the consistency coefficient on temperature can be expressed by the following equation.

$$k = 399.2 \exp(-0.081T) \tag{24}$$

Figure 6 shows the dependence of the consistency coefficient on temperature.

The dependence of oil viscosity on the content of asphaltenes (% wt.) in oil using experimental data is expressed by the formula (**Figure 7**).

$$\eta_{eff} = \eta_0 \left(1 + 0.25\varphi + k_0\varphi^2\right) \tag{25}$$

where φ is the volume fraction of asphaltenes in oil. η_0 is viscosity of oil in the absence of asphaltenes. η_{eff} is effective viscosity of oil. Provided that $\varphi < 10\%$ this expression coincides with the Einstein formula.

In Ref. [17], similar studies were carried out for West Siberian oils for the concentration of asphaltenes in oil from 4 to 72% (wt.). This paper presents experimental studies of the effective viscosity of non-Newtonian oil on the content of asphaltenes at various temperatures. As follows from **Figures 7–9**, the region transitions from Newtonian to non-Newtonian properties of oil as the content of asphalt-resinous substances *I* increases are limited by a stepwise change in oil viscosity for all temperatures. Obviously, this is due to the fact that with an asphaltene content of 38–46% (wt.) in West Siberian oil, there is a stepwise change in the effective viscosity of oil, structural and mechanical strength, transition temperature to the state of a non-Newtonian liquid and molecular weight, which is due to the formation of coagulation structures, aggregates

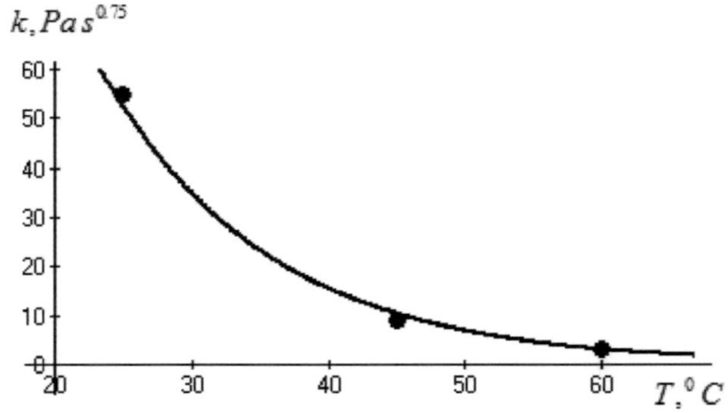

Figure 6.
The dependence of the consistency coefficient on temperature.

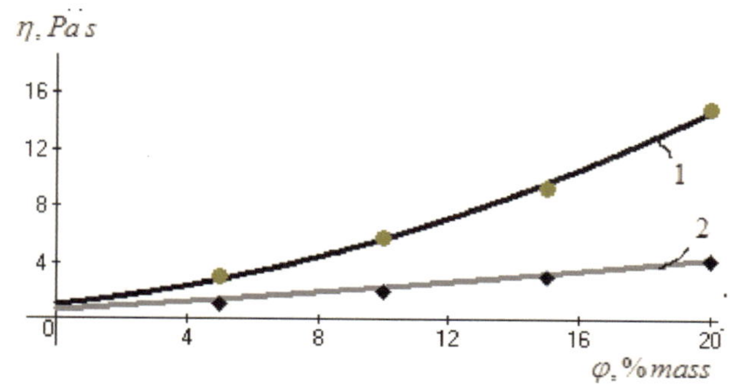

Figure 7.
The dependence of the viscosity of oil on the content of asphaltenes at temperatures: 1 – 25°C, ($k_o = 0.022$); 2 – 45°C, ($k_o = 0.003$).

up to the frame throughout the volume. A stepwise change in viscosity during periods of structure formation and destruction of the structure is a characteristic feature of non-Newtonian oils, which complicates the nature of the description of the entire viscosity and mobility curve of the oil system. The process of formation of coagulation structures is associated with an increase in the probability of interaction and collision of particles with an increase in their concentration in the volume. In the following section, the problems of coagulation, coalescence of droplets and bubbles, and many issues related to the solution of this problem will be discussed in detail.

As follows from **Figure 8**, when structure formation and asphaltene concentration increase, the mobility or fluidity of the oil system decreases, and the fluidity of the system is defined as:

$$\ln \theta = \frac{\ln \eta_{eff\infty}}{\ln \eta_{eff}} \tag{26}$$

where θ is the fluidity of the medium. **Figure 9** shows experimental data on the change in the viscosity of West Siberian oil depending on the content of asphaltenes [18].

The equation describing the experimental data on the viscosity of oil in large intervals of asphaltene content change is presented in the form:

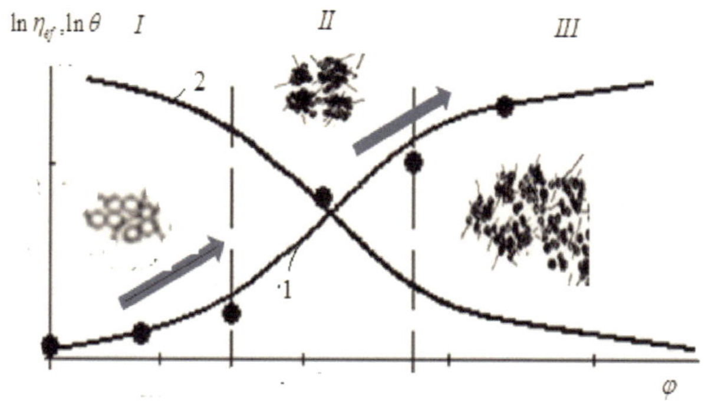

Figure 8.
Characteristic stages of structure formation in oil depending on the content of asphaltenes: I - *oil dispersed system;* II - *area of formation of structures;* III - *structured oil system,* 1 - *viscosity;* 2 - *fluidity.*

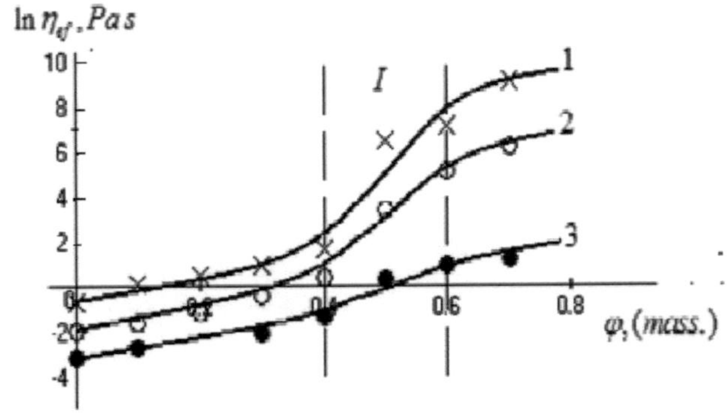

Figure 9.
Dependence of effective viscosity on the content of the dispersed phase of tar-asphaltenes at temperatures: 1 – 84°C; 2 – 112°C; 3 – 144°C. (I–is the region of stepwise change structure formation).

$$\ln \eta_{eff} = \ln \eta_{tff\ 0} + b_0\varphi + b_1\delta(\varphi) + b_2\left(1 - \exp\left(-b_3\varphi^6\right)\right) \tag{27}$$

Here, φ is the mass fraction of asphaltenes in oil, $b_0 - b_3$ is coefficients determined experimentally and depending on temperature, $b_1 = 0.2$ is the maximum value of the delta function, $\delta(\varphi)$ is delta function, determined in the form:

$$\delta(\varphi) = \frac{1}{\exp\left(72.5(\varphi - 0.45)\right) + \exp\left(-72.5(\varphi - 0.45)\right)} \tag{28}$$

μ_0 - initial viscosity,

$$\eta_{eff\ 0} = 2.05 \times 10^{-8} \exp\left(\frac{6075}{T + 273}\right) \tag{29}$$

The value of the delta function characterizes the viscosity jump in the region of structure formation. In particular, the main property of the delta function is the following:

$$\delta(\varphi - \varphi_0) = \begin{cases} 0, \varphi \neq \varphi_0 \\ \infty, \varphi = \varphi_0 \end{cases} \tag{30}$$

The partial approximation expression of the delta function can be represented as:

$$\delta(\varphi) = \frac{1}{\exp\left(b_1(\varphi - \varphi_0)\right) + \exp\left(-b_2(\varphi - \varphi_0)\right)} \tag{31}$$

Here, b_1, b_2 are the coefficients that determine the width of the base of the delta function, φ_0 is the coordinate of the jump center. **Figure 10** shows different kinds of delta functions.

Thus, the use of the delta function makes it possible to describe all the stepwise phenomena occurring during the formation and destruction of structures in non-Newtonian oil. At the same time, satisfactory results are obtained by using a higher-order exponential function, which makes it possible to obtain a soothing effect in the region of the jump.

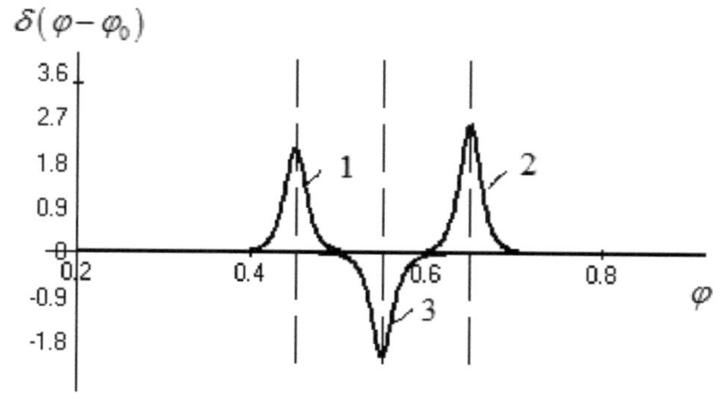

Figure 10.
Delta functions: 1,2 - positive values of the function $\delta(\varphi) > 0$ with centers $\varphi_0 = 0.45, 0.65$; 3 - negative values of the function $\delta(\varphi) < 0$ with the center $\varphi_0 = 0.55$.

The use of aromatic and other solvents partially dissolves asphaltenes, thereby reducing or eliminating the formation of coagulation structures, which improves the rheological properties of oil dispersed media. As follows from **Figure 9**, for this oil, provided that the asphaltene content is less than $\varphi < 0.4$, the formation of coagulation structures is excluded, although for different oil fields there may be other conditions. Analysis of various studies on the effect of asphalt-resin substances on the rheology of non-Newtonian oil of various fields leads to conflicting results, although in all cases an increase in viscosity is observed as a result of structure formation. It should be noted that, in addition to asphaltenes, the rheological properties of oil dispersed system are affected by the content of water and solid-phase in it.

5. Nonlinear equation of filtration of heavy oils in porous media

Oil structured systems containing coagulation structures of crystals of high molecular weight paraffin and asphaltene particles and forming a chain or, in the limiting case, a continuous network (framework), acquire the ability to flow only after the destruction of this network at $\tau >> \tau_0$ (where τ_0 is the yield strength), and small external stresses produce elastic deformation mesh or frame. The interaction of asphaltene particles is accompanied by the creation of sufficiently strong aggregates of a coagulation nature, and above all, doublets and triplets, due to the Brownian diffusion motion of individual particles.

These structures disintegrate into individual particles as a result of the fragmentation of aggregates under the action of shear flow, and the equilibrium shifts towards the formation of individual particles as the shear rate increases. The frequency of collisions of two asphaltene particles in the volume as a result of Brownian diffusion is determined by the following expression [3, 19].

$$\omega = 4\pi(D_1 + D_2)(R_1 + R_2)N_0 \tag{32}$$

The formation of aggregates from asphaltene particles also occurs in pipelines with an intense turbulent flow of oil. The frequency of particle collisions in an isotropic turbulent flow of oil in pipes is determined by the turbulence parameters, the capture coefficient, and the physicochemical properties of oil and asphaltenes. In refs. [2, 3, 19], an expression for the frequency of coagulation and fragmentation of particles in an isotropic turbulent flow due to turbulent diffusion of particles is given in the form:

$$\omega(a) = C_{01} N_0 a^3 \left(\frac{\varepsilon_R}{\nu_c}\right)^{\frac{1}{2}} \exp\left[-C_{02} \frac{\sigma}{(\nu_c \varepsilon_R)^{\frac{1}{2}} a \rho_c}\right] \tag{33}$$

It follows from this expression that the higher the oil viscosity, the lower the frequency of collisions and the lower the probability of formation of coagulation aggregates. Asphaltene particles become larger as a result of coagulation, reach the maximum size of an unstable aggregate, after which their crushing begins. It can be assumed that under the action of hydrodynamic forces, all bonds between particles in an aggregate are stretched to a critical value, as a result of which this aggregate primarily breaks up into smaller aggregates, and then secondary, tertiary etc. decomposition occurs up to a single particle. The destruction of coagulation structures formed by paraffins and asphaltenes is characterized by the fact that after applying a certain load for oil injection, no immediate destruction is observed. In the limiting case of infinite shear rate $\tau > > \tau_0$, aggregates can be completely destroyed up to a single particle, and the flow of such oils or oil emulsions can be considered as the flow of Newtonian fluids. Since the effective viscosity $\eta_{eff}(T, \tau)$ depends on the temperature and on the shear rate (pressure gradient), additional energy will be required to destroy the structure. It should be noted that after the load is removed, the strength of the structure spontaneously recovers completely and the viscosity reaches its maximum value, that is, a certain thixotropy is observed. Thixotropic properties are most typical for structured nonstationary dispersed petroleum systems characterized by aggregative instability and accompanied by aggregation and coagulation of dispersed asphaltene particles. It should be noted that oils and oil products with a high content of asphalt-resinous substances are characterized by thixotropic properties, which are usually described by the Herschel-Bulkley equation and the Ostwald-de Ville power-law, and at high shear rates by the Bingham equation, which is the main equation for describing heavy oils. The equations for the thixotropic behavior of such systems can be represented by various dependencies of the shear stress on time. At the same time, it should be noted that despite the large number of publications offering various approaches to the rheology of structured oils, there is still no satisfactory theory. Linking the rheological coefficients with the structural rheological properties of oil, namely, the forces of interaction between asphaltene particles, the structure, and size of aggregates.

Anomalous viscous-plastic oils differ in their properties from ordinary oils and their rheological description obeys the laws of flow of non-Newtonian Bingham fluids.

$$\tau = \tau_0 + \eta_{eff} \dot{\gamma} \tag{34}$$

From the Newtonian equation ($\eta j \tau\_\gamma$) and taking into account Eq. (34), we obtain:

$$\eta = \eta_{eff} + \frac{\tau_0}{\dot{\gamma}} = \eta_{eff}\left(\frac{\tau_0}{\eta_{eff}\dot{\gamma}} + 1\right), \tag{35}$$

Determining from expression (34)

$$\eta_{eff}\dot{\gamma} = \tau - \tau_0 \tag{36}$$

we finally obtain an expression for the effective viscosity in the form:

$$\eta = \eta_{eff} + \frac{\tau_0}{\dot{\gamma}} = \eta_{eff}\frac{\tau}{\tau - \tau_0}, \tag{37}$$

from which it follows that with increasing τ the value of μ decreases and in the limit at $\tau >> \tau_0$, the value of $\eta \to \eta_{eff}$, corresponding to a system with a completely destroyed structure $\tau_0/_\tau \to 0$. Thus, the viscosity of a structured system in the course of flow under the action of increasing shear stress changes from η_0, corresponding to an undamaged structure, to η_∞, characteristic of a completely destroyed structure. Substituting Eq. (37) into expression (34), we obtain a nonlinear filtration equation for a structured oil system [3, 4].

$$V = -\frac{k_p}{\eta_{eff}}\left(1 - \frac{\tau_0}{\tau}\right)\frac{\partial P}{\partial x} \tag{38}$$

when $\tau >> \tau_0$ this expression goes into the usual Darcy equation for unstructured oil. An analysis of experimental data on the filtration of non-Newtonian oils made it possible to approximate the ratio $\tau_0/_\tau$ in the form:

$$\ln\frac{\tau}{\tau_0} = \alpha\left(\frac{\mathrm{grad}\,P}{(\mathrm{grad}\,P)_0}\right)^{\mathrm{n}} \tag{39}$$

Here, $\alpha-$ is the coefficient determined on the basis of experimental data, n is the exponent. The rheological equation can be written in the form (**Figure 11**)

$$\tau = \tau_0 \exp\left(\alpha(\dot{\gamma}/\dot{\gamma}_0)^n\right) \tag{40}$$

Expression (40) can be considered as a new rheological equation describing the viscoplastic flow of non-Newtonian oils.

Obviously, the index n, depending on the temperature and properties of the porous layer, characterizes the complete destruction of the structure. To date, many concepts

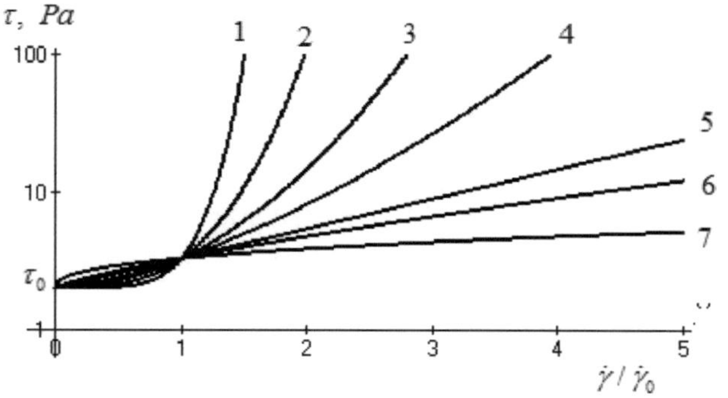

Figure 11.
Dependence of shear stress on shear rate at n*, equal to: 1–5.0; 2–3.0; 3–2.0; 4–1.5; 5–1.0; 6–0.8; 7–0.4.*

α_2	K_2	T^0C	$(\mathbf{grad\,P})_0$
0.08	0.000033	24	0.0140
0.04	0.0000895	50	0.00625
0.02	0.000245	80	0.00470

The dependence of the initial pressure gradient on temperature is given as: $(\mathrm{grad}\,P)_0 = 2.197 \times 10^{-4} + \frac{0.3275}{T}$.

Table 2.
Coefficients of rheological models included in Eq. (41) depending on the temperature.

and models have been put forward to describe the shear flow of oil dispersed systems, resulting in a wide variety of rheological dependences of effective viscosity on shear stress τ and shear rate $\dot{\gamma}$.

6. Rheological models of oil filtration

Using the experimental data of Ref. [18] and Eqs. (38) and (40), we represent the filtration rate in the following form:

$$V = K_2(T)\left(1 - \exp\left(-\alpha_2(T)(z/z_0)^6\right)\right)z \tag{41}$$

where $\alpha_2 = 0.1422\exp(-0.0247T)$, $K_2(T) = 1.4 \times 10^{-5}\exp(0.0364T)$, $z = \mathrm{grad\,P}$, $z_0 = (\mathrm{grad\,P})_0$, $K_2 = k/_{\eta^*}$ are oil mobility. The high value of the exponent is explained by a sharp drop in viscosity during the destruction of the formed structure. **Figure 12** below shows a comparison of the calculated Eq. (41) and experimental values of the filtration rate of heavy oils at different temperatures for various West Siberian fields [18].

The **Table 2** shows the values of the coefficients included in Eq. (41) depending on the temperature.

7. Drag coefficient of particles in a non-Newtonian fluid

Heavy oils differ from ordinary liquids in that their viscosity changes with increasing shear rate, and the shear rate index n characterizes the degree of non-Newtonian behavior of the material. At $n < 1$, liquids exhibit viscous-plastic properties, and at $n > 1$, dilatant properties. These conditions impose special conditions on the description of the hydrodynamics of the flow of non-Newtonian fluids and are reflected in the description of the drag coefficients of particles in such fluids.

Theoretical and experimental studies of hydrodynamics and the drag coefficient of solid particles, drops, and bubbles in a Newtonian fluid are given in Ref. [3], in a power-law non-Newtonian fluid are given in refs. [20–26]. Using asymptotic methods, a formula was obtained in [27, 28] for calculating the drag coefficient of a bubble in a power-law non-Newtonian liquid ($n < 1$) for small numbers $\mathrm{Re}_t < 1$.

$$C_D = \frac{16}{2^n\,\mathrm{Re}_t}X(n), \qquad X(n) = 2^{n-1}3^{\frac{n-3}{2}}\frac{13 + 4n - 8n^2}{(2n+1)(n+2)} \tag{42}$$

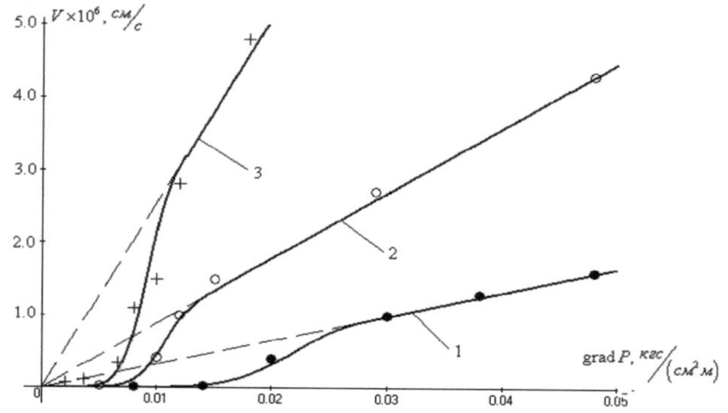

Figure 12.
Change in filtration rate of abnormal oils at different temperatures: 1 - T = 24 C, 2 - T = 50 C;, and 3 - T = 80 C.

where $X(n)$—is a parameter that characterizes the rheological properties of the flow and depends on the exponent n, $\mathrm{Re}_t = \frac{d_T^n \rho_c V^{2-n}}{k_0}$. Calculations using formula (42) show that for pseudo-plastic fluids the drag coefficient is higher, and for dilatant fluids, it is lower than the corresponding drag coefficients, when a Newtonian fluid flows around a bubble. In addition to expression (42), the authors of refs. [27, 28] present the experimental values and various dependencies $X(n)$ on n.

$$X(n) = 2^n 3^{\frac{n-3}{2}}[1 - 3.83(n-1)],\ 0.7 \le n \le 1;$$
$$X(n) = \frac{2}{3}\left(3\gamma^2\right)^{\frac{n-3}{2}} \frac{13 + 4n - 8n^2}{(2n+1)(n+2)},\ \gamma > 10; \tag{43}$$
$$X(n) = 3^{-\frac{n+3}{2}}\left[\frac{2(2n+1)(2-n)}{n^2}\right]^2; \qquad X(n) = 2^n 3^{\frac{n-3}{2}} \frac{1 + 7n - 5n^2}{n(n+2)},\ n < 1.$$

The second formula (43) characterizes the behavior of drops in non-Newtonian fluid. The above expressions, obtained theoretically, do not allow solving this problem for the general case.

A satisfactory dependence of $X(n)$ on n for solid particles in a non-Newtonian liquid for a sufficiently large region of variation $0.1 \le n \le 1.8$ using experimental data [27, 28] can be represented as:

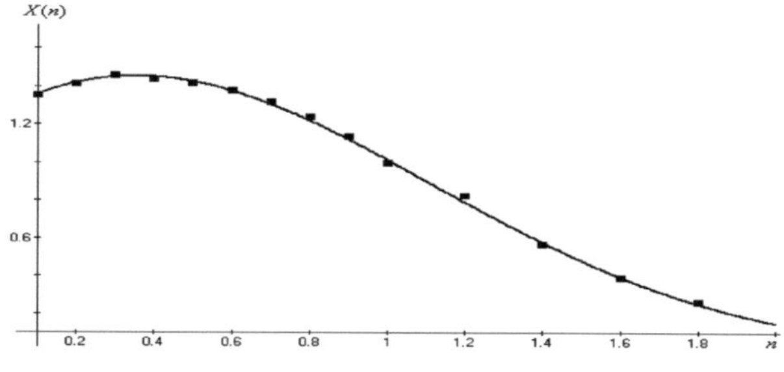

Figure 13.
Dependence $X(n)$ on n.

$$X(n) = \frac{7}{450}(5n+9)^2 \exp\left[-n^{\frac{3}{2}}\left(1+\frac{n^{\frac{3}{2}}}{10}\right)\right] \tag{44}$$

It follows from formula (44) that at $n = 1$, the value of $X(n) \approx 1$. **Figure 13** shows a comparison of the calculated values with the experimental values.

In refs. [27, 28], various formulas are also given that characterize the deformation of bubbles in a non-Newtonian fluid.

$$\begin{aligned} & Y_K = \frac{b_0}{a_0} = 0.0628\delta^{0.46}, \quad 20 \le \delta \le 100, \\ & Y_K = 1.4, \quad \delta \le 4, \\ & Y_K = 6.17\delta^{-1.07}, \quad 4 \le \delta \le 20, \\ & \delta = \mathrm{Re}_t \mathrm{Mo}_t^{0.078}, \quad \mathrm{Mo}_n = \mathrm{We}^{n+2}\mathrm{Fr}^{2-3n}\,\mathrm{Re}_t^{-4}, \quad 0.64 \le n \le 0.9 \end{aligned} \tag{45}$$

where a_0, b_0 are the major and minor axes of the ellipsoid.

For small values of the number $5 \le \mathrm{Re}_d \le 25$, the drag coefficient of a bubble in a non-Newtonian liquid, according to the works in refs. [2, 27, 29], can be determined in the form:

$$C_{DG} = \frac{K(n)}{\mathrm{Re}_t} \tag{46}$$

Here $K(n)-$ is a coefficient depending on n. Using the experimental data given in refs. [27, 28], the following formula is proposed in this work:

$$K(n) = 30.6\frac{\exp(2.53n)}{\left(n^{\frac{6}{7}}+1\right)^3} \tag{47}$$

As follows from **Figure 14**, the coefficient $K(n)$ varies significantly depending on $0 \le n \le 1$ and, passing through a minimum at $n = 1$, tends to $K(n) \approx 48$, which characterizes the drag coefficient of a bubble in a Newtonian fluid, proposed in Ref. [29] for small numbers Re_d.

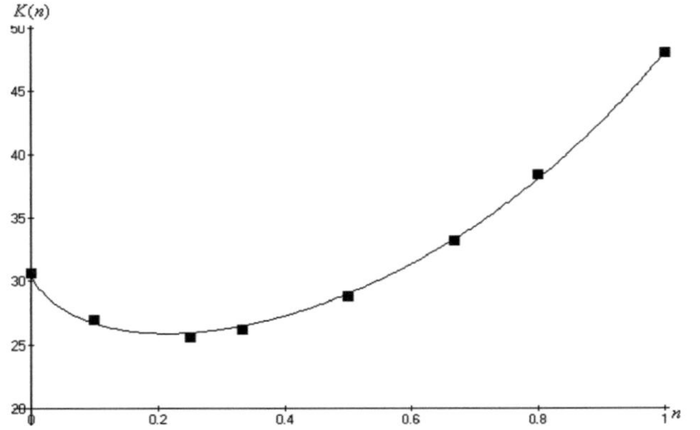

Figure 14.
Dependence of $K(n)$ on n.

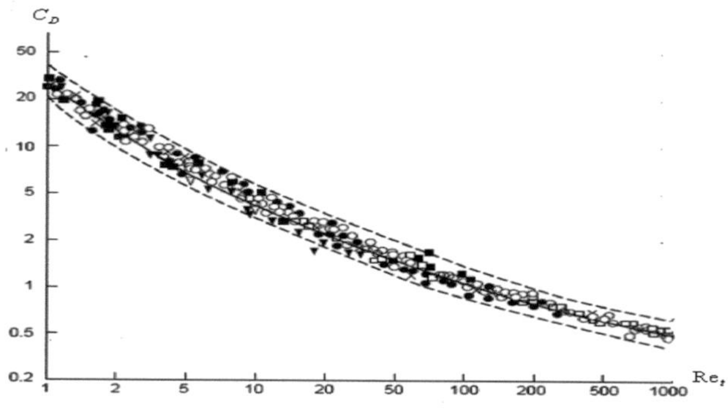

Figure 15.
Drag coefficient of solid particles in non-Newtonian liquid (solid line - drag coefficient in Newtonian fluid and dotted lines correspond to 30% of the experimental data scatter threshold) [30].

A lot of experimental data for the drag coefficient of particles in a non-Newtonian fluid collected from the literature for $Re_d < 1000$, is given in refs. [30, 31] and in **Figure 15**.

In this paper, using experimental data from Ref. [22], we propose an equation for determining the drag coefficient of solid particles in a power-law non-Newtonian fluid for a wide range of number variation $0.1 < Re_t < 10^5$.

$$C_D = \frac{24X(n, Re_t)}{Re_t}\left[1 + Re_t^{\frac{2}{3}}\xi(Re_t)\right] + 0.3\left(1 - \exp\left(-0.2 \times 10^{-19} Re_t^4\right)\right)$$
$$X(n, Re_t) = \frac{5 \cdot 6^{n-1}}{1+4n}\left[1 + (1-n)^{\frac{1}{3}} \frac{Re_t^{\frac{3}{2}}}{1 + (1-n)^{\frac{1}{3}} Re_t^{\frac{3}{2}}}\right], \quad 0.8 < n \le 1, \quad 0.1 \le Re_t \le 10^5$$
$$\xi(Re_t) = \frac{Re_t^2}{5 + 8 Re_t^2} + \frac{Re_t}{6.8 \times 10^6 Re_t^{-\frac{5}{4}} + 32 Re_t^{\frac{2}{3}} + 952.8 \times 10^{-13} Re_t^{\frac{10}{3}}} \tag{48}$$

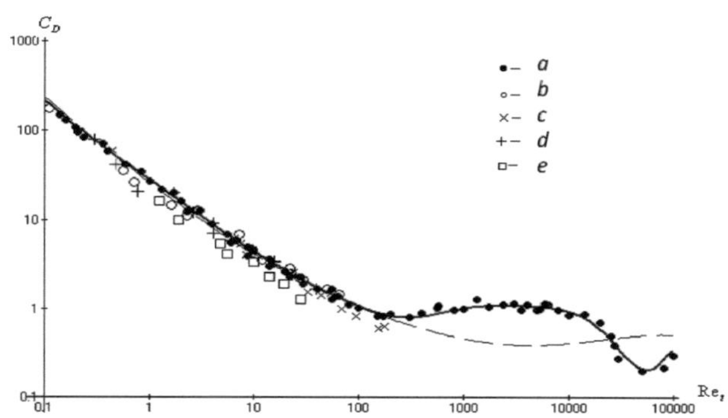

Figure 16.
The dependence of the drag coefficient of a solid particle in a non-Newtonian fluid on the number Re_d according to the data of various authors (dashed line is the drag coefficient of a particle in a Newtonian fluid): $a - n = 0.84 - 0.86$ [32]; $b - n = 0.75 - 0.90$ [32]; $c - n = 0.75 - 0.92$ [33]; $d - n = 0.56 - 0.75$ [34]; $e - n = 0.73$ [35].

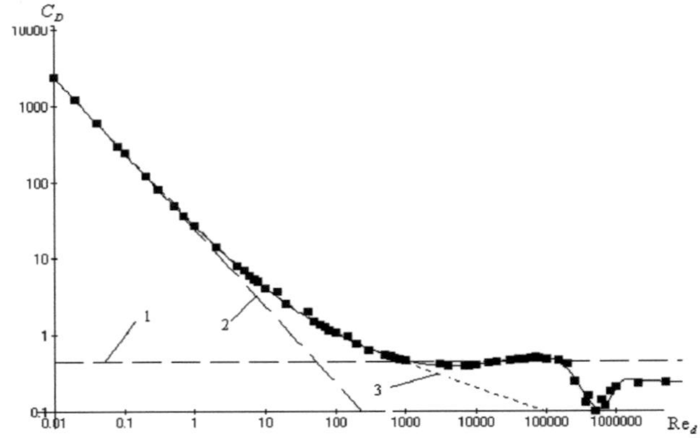

Figure 17.
The drag coefficient for a solid spherical particle.

For values $\mathrm{Re}_t < 1000$, this equation simplifies to the form:

$$C_D \approx \frac{24X(n, \mathrm{Re}_t)}{\mathrm{Re}_t}\left(1 + 0.125\,\mathrm{Re}_t^{\frac{2}{3}} + 2.51 \times 10^{-5}\,\mathrm{Re}_t^2\right) \tag{49}$$

Figure 16 compares the calculated values of the drag coefficient according to Eq. (49) with the experimental data given in refs. [31–35]. As follows from this figure and calculations using the formula in Eq. (49), decreasing the drag coefficient of solid particles in viscous-plastic liquids is less than in Newtonian ones.

The drag coefficient of solid spherical particles in a Newtonian fluid is shown in **Figure 17** [1, 3]. Comparing **Figures 16** and **17** in the region of the drag crisis, it should be noted that when a non-Newtonian flow flows around a solid particle, the drag crisis occurs much earlier than in a Newtonian fluid.

$$1 - C_D \approx 0.44; 2 - C_D = \frac{24}{\mathrm{Re}_d}; 3 - C_D = \frac{24}{\mathrm{Re}_d}\left(1 + 0.15\,\mathrm{Re}_d^{0.687}\right). \tag{50}$$

8. Free settling of particles in heavy oil

The motion of a single particle in a force field with a slow flow of the medium, taking into account the added mass, the weight force, corrected for the Archimedes force and the resistance force, is described by the Eq. (3)

$$\frac{d\mathbf{V_p}}{dt} = \frac{\Delta\rho}{2\rho_d + \rho_c}\mathbf{g} - \frac{3}{4}C_D\frac{\rho_c}{2\rho_d + \rho_c}\left|\mathbf{V_p} - \mathbf{U}\right|\left(\mathbf{V_p} - \mathbf{U}\right) \tag{51}$$

In the steady-state, this equation for a Newtonian fluid, as a result of the balance of forces acting on a particle and for small numbers Re, is transformed into a simpler form:

$$\frac{4}{3}\pi a^3 \Delta\rho g = \rho_C V_s^2 \pi a^2 C_D \tag{52}$$

From this equation provided that $C_D = 24/\mathrm{Re}$ we determine the deposition rate in small numbers $\mathrm{Re}_d < < 1$ in the form of the stokes model $V_s = \frac{1}{18}\frac{\Delta\rho a^2 g}{\eta_c}$.

Figure 18 shows characteristic curves and experimental data on the settling rate. Given expressions for a non-Newtonian fluid are as follows:

$$C_D = \frac{24}{2^n \operatorname{Re}} X(n), \qquad X(n) = 2^{n-1} 3^{\frac{n-3}{2}} \frac{13 + 4n - 8n^2}{(2n+1)(n+2)}, \text{ and } \operatorname{Re}_t = \frac{a^n \rho_c V^{2-n}}{k_0}. \tag{53}$$

Let us rewrite Eq. (52) for a power-law non-Newtonian fluid, it will be represented as:

$$\frac{4}{3} \frac{a \Delta \rho g}{\rho_C} \frac{\operatorname{Re}}{24 X_n} = V_s^2 \tag{54}$$

$$X_n = 2^{-1} 3^{\frac{n-3}{2}} \frac{13 + 4n - 8n^2}{(2n+1)(n+2)}$$

Transforming Eq. (54), we finally obtain an expression for the settling rate of a single solid particle in a non-Newtonian fluid flow.

$$V_s = \left(\frac{1}{18} \frac{\Delta \rho g}{k_0 X_n} \right)^{1/n} a^{(n+1)/n} \tag{55}$$

It should be noted that if $n = 1$, then $X_n = 1$ and expression (55) also turns into the Stokes equation for the deposition of a solid particle in a Newtonian fluid at small numbers Re.

$$V_s = \frac{1}{18} \frac{\Delta \rho g a^2}{\eta_C} \tag{56}$$

Obviously, an important role in the migration and sedimentation of particles belongs to the resistance forces, depending on the number $\operatorname{Re}_d = {}^{Ua}/_{\nu_c}$, shape and size of particles, on the physicochemical properties of particles, and the medium.

Numerical calculations using the above formulas show that for fine-dispersed particles of asphaltenes and paraffins, with increasing particle size, the degree of

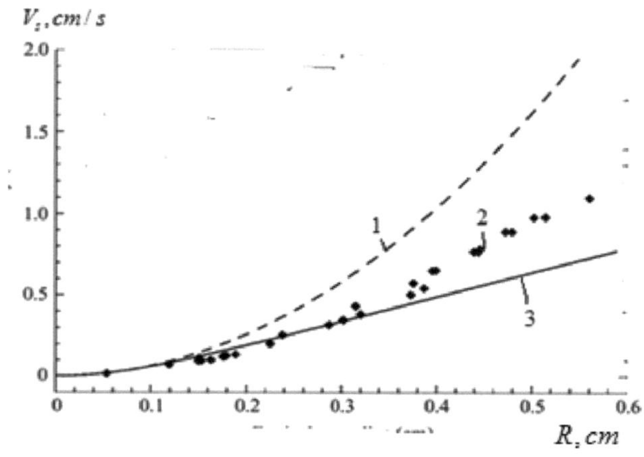

Figure 18.
Settling velocity profiles depending on particle radius: 1 - Hadamard-Rybczynski equation, 2 - experiment, and 3 - Eq. (52).

entrainment of particles by a pulsating medium decreases, and fine-dispersed particles react to turbulent pulsations of the medium, perform under their influence a pulsating motion relative to the moles of the carrier phase and random motion in all directions due to turbulent diffusion. These parameters are very important factors that determine the degree of migration and settling of particles in a turbulent flow. It should be noted that for Stokes spherical solid particles at $\mathrm{Re}_d < 1$, the rate of their settling from the volume is determined as:

$$V_s = \tau_p g \tag{57}$$

For a viscous-plastic fluid obeying the Ostwald–de Ville model $\tau = \eta_c \dot{\gamma}^n$, Eq. (56) can be written as:

$$V_p = \frac{1}{18} \frac{\Delta \rho g a^2}{\tau / \dot{\gamma}^n} \tag{58}$$

with a relaxation time equal to

$$\tau_P = \frac{\rho_d a^2 \dot{\gamma}^n}{18 \tau} \tag{59}$$

For a viscous flow of a power-law fluid, we obtain the expression for the relaxation time in the form:

$$\tau_P = \left(\frac{1}{18} \frac{\Delta \rho g}{k_0 X_n} \right)^{1/n} a^{(n+1)/n} \tag{60}$$

Using the experimental data for the deposition of glass particles in a non-Newtonian liquid [36], it can be noted that the deposition rate dependence is linear only at low shear rates.

A somewhat different formula for the deposition of single particles in a non-Newtonian liquid was proposed in Ref. [37], which can be obtained from (58) at $X_n = 2 - n, \quad 0.4 < n < 1.2$

$$V_p = \left(\frac{\Delta \rho g}{18 k_0 (2 - n)} \right)^{1/n} a^{\frac{n+1}{n}} \tag{61}$$

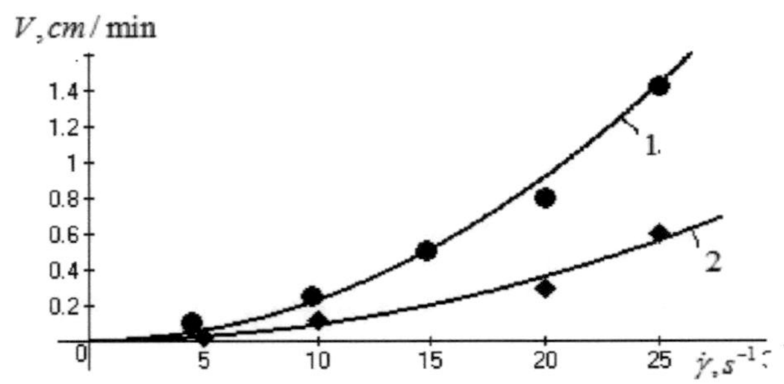

Figure 19.
Dependence of the particle settling rate in a non-Newtonian liquid on stress and shear rate for particles of size $a = 275\,\mu m$, $1 - \tau = 8.5\,Pa$; $2 - \tau = 15\,Pa$.

Using the experimental data on the settling of solid particles in a non-Newtonian fluid [38] in **Figure 19** shows a comparative characteristic according to the model (58).

For large numbers of Re, expression (52) can be written as:

$$\mathrm{Re}^{2}C_{D}^{2-n}(n,\ \mathrm{Re}) = \left(\frac{8}{3}\Delta\rho g\right)^{2-n}\rho_{C}^{n}\frac{a^{n+2}}{k^{2}} \tag{62}$$

Expression (62) does not allow an analytical determination of the settling rate and is solved only numerically.

9. Flow of heavy oil in pipes with settling

During the flow of heavy oil, asphalt-resinous substances and particles of the solid phase are deposited on the inner surface of the pipes. As a result of the deposition of particles on the surface of the pipe, a layer of deposits is formed, which continuously grows and is compacted by the normal component of pressure. The quantitative content of asphaltenes and resins, due to their high energy of adhesion and cohesion, significantly affect the rheological properties of heavy oils and the formation of a dense layer on the surface. The high ability to form coagulation structures and aggregates contributes to the deterioration of the hydrodynamic characteristics of the flow of non-Newtonian oils up to zero flow velocity due to an increase in the effective viscosity of the medium. The mechanism of sedimentation of particles from heavy oil can be carried out by diffusion (for a vertical surface), gravitational (for a horizontal surface), and diffusion-gravitational way. The formation of a dense layer on the inner surface of the pipe has a significant effect on the transfer of mass and heat, and on the hydrodynamics and rheological structure of the flow.

The hydrodynamic equation for the flow of viscous-plastic oil in a pipe, described by the Bingham rheological equation $\tau = \tau_0 + \eta_C\dot{\gamma}$ for the stationary case in cylindrical coordinates, will be represented as:

$$\frac{1}{r}\frac{\partial}{\partial r}\left(r\frac{\partial V}{\partial r}\right) = \frac{1}{\eta_C}\frac{\partial P}{\partial x} \tag{63}$$

The boundary conditions for the flow of viscous-plastic oil in the pipe are (**Figure 20**)

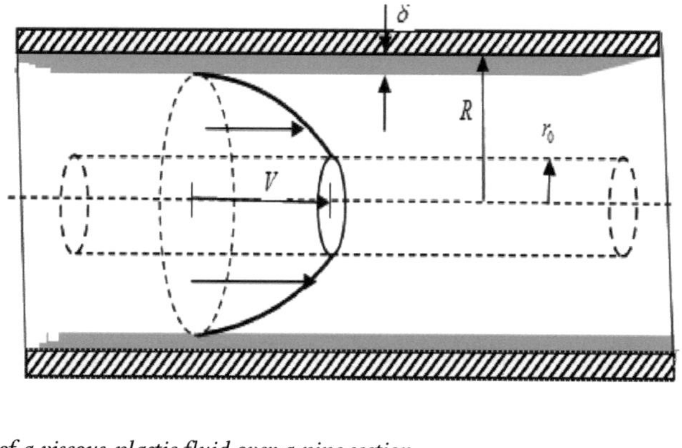

Figure 20.
Velocity distribution of a viscous-plastic fluid over a pipe section.

$$r = R,\ \ \tau < \tau_0,\ \ \tau = \tau_0 + \eta_C \frac{\partial V}{\partial r} = 0,\ \ V = 0 \tag{64}$$

$$r = R,\ \ \tau > \tau_0,\ \ \tau = \tau_0 + \eta_C \frac{\partial V}{\partial r} \neq 0,\ \ V = 0 \tag{65}$$

Integrating condition (64), we have:

$$V = -\frac{\tau_0}{\eta_C} R \tag{66}$$

Integrating condition (65), we obtain:

$$V = \frac{\tau_0 R}{\eta_C}\left(1 - \frac{\tau}{\tau_0}\right) \tag{67}$$

Integrating expression (63) twice, we obtain:

$$V = -\frac{1}{4\eta}\frac{\partial P}{\partial x} r^2 + A \tag{68}$$

Here, A is the integration coefficient equal to for the first boundary condition (64)

$$A = \frac{1}{4\eta_C}\frac{\partial P}{\partial x} R^2 - \frac{\tau_0}{\eta_C} R \tag{69}$$

and for the second boundary condition (3.43)

$$A = \frac{1}{4\eta}\frac{\partial P}{\partial x} R^2 - \frac{\tau_0 R}{\eta}\left(1 - \frac{\tau}{\tau_0}\right) \tag{70}$$

As a result of solving the hydrodynamic Eq. (63) for a viscous-plastic fluid described by the expression in a pipe with boundary conditions (64) and (68) and assuming that, taking into account (68)–(70), in the simplest case, we obtain the following distribution for the flow velocity (**Figure 20**)

$$\begin{aligned} V(r) &= \frac{\Delta P R^2}{4\eta_C l}\left(1 - \frac{2l}{R}\frac{\tau_0}{\Delta P}\right),\ \ r < r_0 \\ V(r) &= \frac{\Delta P R^2}{4\eta_C l}\left[1 - \frac{r^2}{R^2} - \frac{4l}{R}\frac{\tau_0}{\Delta P}\left(1 - \frac{r}{r_0}\right)\right],\ \ r_0 \le r < R \end{aligned} \tag{71}$$

where l is the pipe length and r_0 is the plug radius. For the case of the formation of a dense layer of particles on the inner surface of the pipe with a thickness of δ, which is typical for the flow of viscous oil and, assuming that $R = R_0\beta,\ \beta = 1 - \delta/R_0$ (δ is the thickness of the sediment layer), then these expressions (71) for the quasi-stationary case will be presented as:

$$\begin{aligned} V(r) &= \frac{\Delta P R_0^2\beta^2}{4\eta_C l}\left(1 - \frac{2l}{R_0\beta}\frac{\tau_0}{\Delta P}\right),\ \ r < r_0 \\ V(r) &= \frac{\Delta P R_0^2\beta^2}{4\eta_C l}\left[1 - \frac{r^2}{R_0^2\beta^2} - \frac{4l}{R_0\beta}\frac{\tau_0}{\Delta P}\left(1 - \frac{r}{r_0}\right)\right],\ \ r_0 \le r < R_0\beta \end{aligned} \tag{72}$$

It should be noted that if $\beta = 1$, then the thickness of the deposits is absent, and if $\beta = 0$, then $\delta = R_0$; there is a complete blockage of the pipe and the flow velocity $V(r) \to 0$, that is, carrying capacity is reduced to almost zero.

This distribution of velocities in hydrodynamics is called the "structural regime of motion." The volumetric flow rate of viscous-plastic oil over the pipe section is determined as:

$$Q = \frac{\pi R^4 \Delta P}{8l\eta_c}\left(1 - \frac{4}{3}\frac{r_0}{R} + \frac{1}{3}\frac{r_0^4}{R^4}\right) \tag{73}$$

As a result of the formation of a dense layer on the inner surface of the pipe, assuming that $R = R_0\beta$ you can write:

$$Q = \frac{\pi R_0^3 \beta^3 \Delta P}{8l\eta_c}\left(1 - \frac{4}{3}\frac{r_0}{R_0\beta} + \frac{1}{3}\frac{r_0^4}{R_0^4\beta^4}\right) \tag{74}$$

As follows from this equation, with an increase in the thickness of deposits on the inner surface of the pipe, which corresponds to a decrease in β, the volume flow of oil decreases. Thus, with a complete blockage of the pipe section, the fluid flow also tends to zero. For a Bingham fluid, the conditions for the transition from a structural flow to a turbulent regime are defined as:

$$\mathrm{Re}_{cr} = \frac{1 - 4\alpha + \alpha^4}{24\alpha}\mathrm{He}, \ \mathrm{He} = 16800\frac{\alpha}{(1-\alpha)^2} \tag{75}$$

or you can write

$$\mathrm{Re}_{cr} = 700\frac{2 + (1+\alpha)^2}{1-\alpha} \tag{76}$$

Here, $\mathrm{Re}_{cr} = V_{cr}d/\nu_c$ is the critical Reynolds number, $\mathrm{He} = \tau_0 d^2/\nu_c$ is the Hedstrom number, and $\alpha = \tau_0/\tau_R$, τ_R is the shear stress on the pipe wall with a diameter equal to d. In the presence of a dense layer on the inner surface of the pipe, one can write $\mathrm{He} = \mathrm{He}_0\beta^2$, that is, as the thickness of the deposits grows, the intensity of turbulence decreases. For some oils, power-law of flow is observed $\tau = k\dot{\gamma}^n$, for which one can write:

$$\mathrm{Re} = \frac{8}{k_0}\left(\frac{n}{6n+2}\right)^n \rho_c d^n V^{2-n} = \mathrm{Re}_0\beta^{3n-4} \tag{77}$$

Thus, the transition from a structural flow regime to a turbulent one occurs at certain values of stress τ and dimensionless deposit thickness β, and, if $n > 1$ the turbulence intensity decreases as the thickness increases and, conversely, at $n \leq 1$– the turbulence intensity increases.

10. Practical ways to improve rheological properties at heavy oil refining

Heavy oils, due to the high content of heavy components (65–70% fraction >350^0C boiling) exhibit viscous-plastic properties of non-Newtonian oils. Currently, there are no special technologies for the processing of heavy oils, with the exception of some works [39, 40], since their processing is limited by the problems of oil transport in pipes, pumps, heat exchangers, and other equipment. The viscosity of heavy oils, which determines their fluidity, depends on the content of asphalt-resinous substances, paraffins, water-in-oil, and other factors [41–44]. When processing heavy oils at operating plants at relatively low temperatures, the effective viscosity of the oil increases, which helps to reduce its fluidity and, in general, the productivity of the process. The decrease in the viscosity of heavy oils, which ensures their transportability, depends on two factors: an increase in its temperature and dilution with light components.

In this regard, the main factor for improving the rheological properties of oil and the conditions for its processing when solving practical problems is to increase the temperature of the oil stock at the inlet to the plant and its partial dilution by creating a recirculation of a lighter component for mixing with crude oil (**Figure 21**). This intensification of the processing process allows us to solve both problems simultaneously. The process of primary oil refining is carried out by its preliminary heating to a certain temperature with its intermediate purification from water and salts in the dehydrator 4, contained in the oil, and further separation in the mass-exchange distillation column 5 (**Figure 20**).

The unit used Kazakhstan heavy paraffinic oil with high viscosity (**Figure 22**).

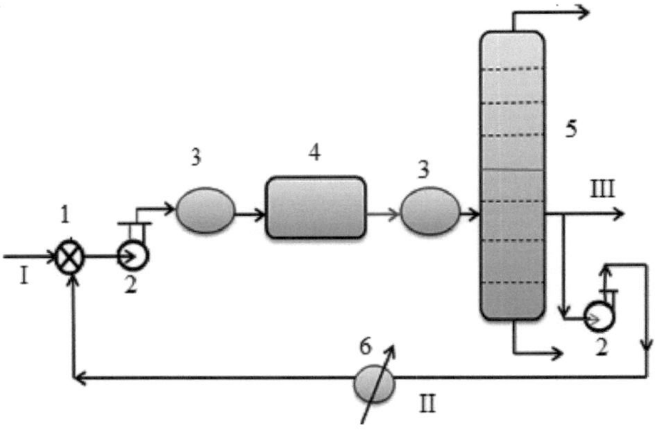

Figure 21.
Scheme of primary oil refining with recirculation: 1 - mixer, 2 - pump, 3 - system of heat exchangers, 4-dehydrator, 5 - distillation column, and 6 -refrigerator; I - crude oil, II -recirculation line, and III - fraction >240.

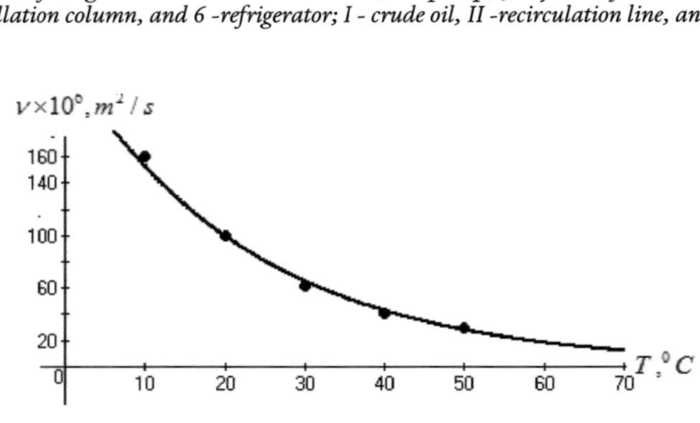

Figure 22.
Temperature dependence of crude oil viscosity.

The dependence of the kinematic viscosity of crude oil on temperature is represented by the expression

$$\nu = 234.05 \exp\left[-0.048T\right] \times 10^{-6}, m^2/s \tag{78}$$

To improve rheological properties and viscosity in order to ensure its transportability, it is proposed to return part of the >240^0C fraction from column 5 back for mixing with crude oil after preliminary cooling to pump 2. The creation of a recirculation system makes it possible to reduce the kinematic viscosity of crude oil by increasing its temperature to 20–25^0C and diluting it with a lighter liquid. Experimental studies have shown a significant effect of the amount and temperature of the >240^0C fraction on the effective viscosity of crude oil (**Figure 23**).

As follows from **Figure 23** with an increase in temperature and the amount of added fraction, the viscosity of the oil decreases. The expression describing the change in the kinematic viscosity of oil from the temperature at various fraction contents >240^0C can be represented as:

$$\nu = 108.95(1 - 2.17\varphi) \exp\left(-0.028T\right) \times 10^{-6} \tag{79}$$

Here, $\varphi = m_R/(m_O + m_R)$–is the mass fraction of the >240^0C fraction in crude oil, and m_O, m_R are the mass flow rates of crude oil and fractions. The viscosity of heavy oil due to dilution with a light fraction is defined as:

$$\nu = \frac{\nu_O(T)}{1 + m\varphi^2} \tag{80}$$

Data of **Figures 23** and **24** show the dependence of the viscosity of oil when it is diluted with a light fraction, calculated by Eq. (80).

As follows from **Figure 23**, the relative fluidity of heavy oil increases, as a result of its dilution with a light fraction, which improves the rheological properties of the oil.

The temperature of a mixture of crude oil diluted with a light fraction >240^0C is defined as:

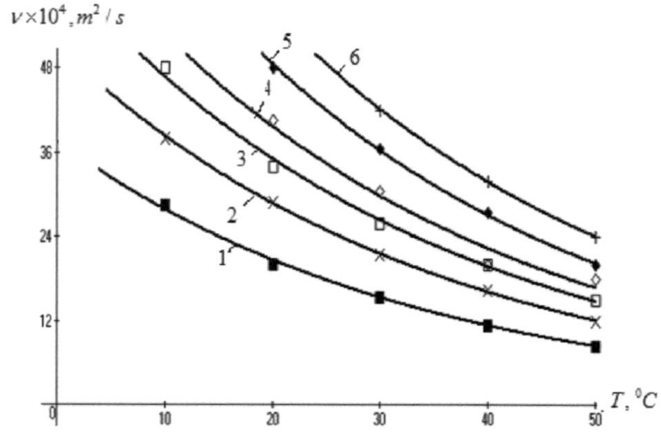

Figure 23.
The dependence of the kinematic viscosity of oil on temperature when it contains fractions >240°C φ (mass fractions), is equal to: 1–0.3; 2–0.25; 3–0.20; 4–0.15; 5–0.10; 6–0.05, $\nu \times 10^4, m^2/s$.

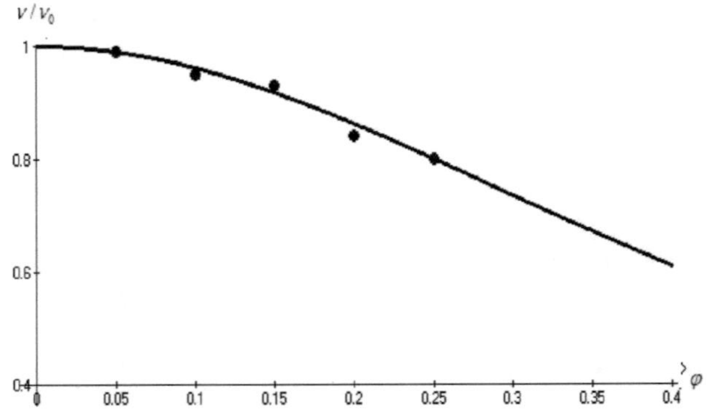

Figure 24.
The dependence of the viscosity of heavy oil on the proportion of the light fraction at a temperature $T = 30°C, m = 3.5$.

$$T = (1 - \varphi)c_{p0}T_0 + \varphi c_{pR}T_R \tag{81}$$

Here T is the temperature of the mixture at the inlet to column 5; c_p is the heat capacity of crude oil and fraction; and T_0, T_R are temperatures of crude oil and fraction.

Recirculation of some lighter, higher temperature, lower viscosity refining fraction for blending with crude oil reduces the viscosity of the oil by raising its temperature and partially diluting it with the lighter fraction (**Figure 24**).

11. Analysis and discussion of results

The main problem of the rheology of heavy oils is the study of the regularity of their behavior under the action of external deforming stresses, taking into account their structure. At the same time, the processes associated with irreversible residual deformations and the flow of various viscous and plastic materials, as well as the phenomena of relaxation of heavy oils, allow us to study the fundamental properties of oil disperse systems characterized by shear stress, effective viscosity, corresponding to a certain disordered structure of the system. The effective viscosity of non-Newtonian oils depends on temperature and pressure, as well as on the shear strain rate, the properties of the disordered structure, the content of asphalt-resin substances and the concentration of the dispersed phase, and determines the degree of oil transportability during its extraction and processing.

Concluding this study, we note the main stages and problems of the rheology of heavy oils considered in this paper:

- A high content of asphalt-resinous, paraffinic substances and solid phases of various types and nature in heavy oils creates a special physical structure that gives them viscous-plastic properties, the description of which is based on the rheological models of Bingham, Hershel-Bulkley, and Ostwald-de Ville model. Such high-viscosity oil systems are formed as a result of the combination of oil emulsions with the presence of water droplets, oil suspensions with the presence of a solid phase in the composition of crude oil and gas suspensions, including gas bubbles, as well as asphalt-resinous and waxy substances dissolved in oil. The complex physical structure of heavy oil predetermines the possibility of the formation of more complex coagulation structures that adversely affect the rheology.

- For the filtration of heavy oils in an anisotropic porous medium, a new nonlinear filtration equation is proposed that takes into account shear stress, elastic limit, and effective viscosity. Analysis and solutions of this equation show a high degree of dependence on the filtration rate and effective viscosity on the pressure gradient. It is noted that the nonlinearity of the rheological model of filtration in porous media is determined by the nature of the hydrodynamic flow and the high content of asphalt-resinous and paraffinic substances in the composition of heavy oil. Solutions of the rheological filtration model make it possible to determine the filtration rate depending on the shear stress or the pressure drop and on the effective viscosity.

- Expressions were proposed for determining the flow rate of heavy oils and the rate of free settling of particles, particulate matter and particle drag coefficients in heavy oils.

- Proposed effective ways to improve the rheological properties of oil during its processing at existing plants, ways to create a recirculation scheme.

The problems associated with the rheology of heavy oils considered in this paper show the complexity, both in theoretical and practical aspects, of solving these problems, which is determined by the course of many interconnected physical phenomena and the high randomness of these processes. The relationship between determinism and randomness in these phenomena gives rise to some problems that lead to the use of empirical models that do not take into account the true physical mechanisms and phenomena. In particular, there is no way to take into account the nature and properties of the formed disordered structures, and the conditions for their formation and destruction in rheological models. All these phenomena are taken into account in rheological models indirectly in the form of introducing the volume fraction of particles into the model in the form of various functions. As a result of this, the same experimental data can be simultaneously described by many empirical models within a given accuracy. Obviously, taking into account these factors will lead to a more complex structure of the rheological model, and therefore, for practical calculations, it is possible to use various empirical models for specific rheological fluids.

Nomenclature

a	is particle diameter
C_D	is drag coefficient of particle
D_M	is coefficient of molecular diffusion
D_{eff}	is coefficient of effective diffusion
d	is diameter of the pipe
k_0	is consistency factor'

k_p is permeability coefficient
m is mass of substance
N is total number of particles per unit volume
n is an indicator of the degree of non-Newtonian fluids
P is total pressure
R is radius of the particle
t is time
T is temperature
U is the average flow velocity
V_p is total particle deposition rate

Greek letters

β is dimensionless thickness of particle deposits
ε_R is specific energy dissipation per unit mass
ε is porosity
η_{eff} is effective viscosity of suspensions
η_c, η_d are dynamic viscosity of the medium and particles
ν_c, ν_d are kinematic viscosity of the medium and particles
ρ_c, ρ_d are density of the medium and particles
ρ_i is concentration of particles
σ_D is coefficient of surface tension
τ is shear stress
τ_0 is yield strength
φ is the volume fraction of particles
ω is frequency of turbulent coalescence
He is the Hedstrom number
Re is the Reynolds number
Wi is the Weissenberg number

Indices

d is a solid particle
c is medium
0 is the initial value

Acknowledgements

This work was supported by the Science Foundation of «SOCAR» under the grant project 13LR - AMEA (05/01/2022) at the Institute of Catalysis and Inorganic Chemistry named after Acad. M.F.Nagiyev.

Author details

Gudret Isfandiyar Kelbaliyev, Dilgam Babir Tagiyev and Manaf Rizvan Manafov*
Institute of Catalysis and Inorganic Chemistry named after academician M. Nagiyev of the National Academy of Sciences of Azerbaijan, Baku, Azerbaijan

*Address all correspondence to: mmanafov@gmail.com

References

[1] Kelbaliev GI, Rasulov SR, Tagiev DB, Mustafayeva GR. Mechanics and Rheology of Petroleum Dispersed Systems. Moscow: Mask Publishing House; 2017. p. 462

[2] Kelbaliyev GI, Tagiyev LB, Rasulov SR. Transport Phenomena in Dispersed Media. Taylor and Francis Group: CRC Press Boca Raton-London-New York; 2019. p. 434

[3] Kelbaliyev GI, Tagiyev DB, Manafov MR. Rheology of structured oil emulsion, in book: nano- and micro- encapsulation - techniques and applications. London: IntechOpen. 2021:1-38. DOI: 10.5772/intechopen.92770

[4] Pal R. Fundamental rheology of disperse systems based on single-particle mechanics. Fluids. 2016;**1**(40). DOI: 10.3390/fluids1040040

[5] Rannigsen HP. Rheology of petroleum fluids. Annual Transaction of the Nordic Rheology Society. 2012;**20**:11-18

[6] Ilyin SO, Larisa A. Strelets basic fundamentals of petroleum rheology and their application for the investigation of crude oils of different natures. Energy & Fuels. 2018;**32**(1):268-278

[7] Kirsanov SV, Matveenko VN. Non-Newtonian Behavior of Structured Systems. Moscow: Technosphere; 2016

[8] Zlobin AA. Study of mechanism of oil magnetic activation in order to protect production wells from wax deposition. Bulletin of PNRPU. Geology. Oil & Gas Engineering & Mining. 2017;**16**(1): 49-63. DOI: 10.15593/2224-9923/2017.1.6

[9] Byron BR, Armstrong RC, Hassager O. Dynamics jf Polymeric Liquids. Vol. 1. John Wiley Sons. Inc; 1987

[10] Barnes HA. A review of the rheology of filled viscoelastic systems. Rheology Reviews. 2003:1-36

[11] Finotello G, De Jeroen S, Vrouwenvelder CR, Padding JT, Buist KA, Jongsma A, et al. Experimental investigation of non Newtonian droplet collisions the role of extensional viscosity. Experiments in Fluids. 2018; **59**(113):112-128. DOI: 10.1007/s00348-018-2568-2

[12] Alabi OO. Validity of Darcy's law in laminar regime. Electronic Journal of Geotechnical Engineering, Bundle A. 2011;**16**:27-40

[13] Klaus Kroy, Isabelle Capron, Madeleine djabourov. Physiqe Thermique, ESPCI. 10, rue Vauquelin, Paris. France. 1999.

[14] Krassimir D. Danov on the viscosity of dilute emulsions. Journal of Colloid And Interphase Science. 2001;**235**: 144-149

[15] Argillier JF, Coustet C, Henau I. Heavy oil rheology as a function of asphaltene and resin content and temperature. In: Society Petroleum Engineers. 2002. DOI: 10.2118/79496-MS

[16] Davarpanah L, Vahabzadeh F, Dermanaki A. Structural study of asphaltenes from iranian heavy crude oil. Oil & Gas Science and Technology - Revue d'IFP Energies nouvelles. 2015; **70**(6):1035-1049

[17] Mukhmedzyanova AA, Buduik VA, Alyabev AS, Khaybullin AA. Influence of temperature and asfaltenes concentration on reological prorerties of disperse systems of the west Siberian oils tar. Bashkir Chemical Journal. 2012; **19**(4):1-4

[18] Sarimeseli A, Kelbaliyev G. Modeling of the break-up particles in developed turbulent flow. Chemical Engineering Science. 2004;**59**:1233-1243

[19] Devlikamov II, Khabibullin ZA, Kabirov MM. Anomalous oil. M: Nedra. 1975:168

[20] Clift KA, Lever DA. Isothermal flow past a blowing sphere. International Journal for Numerical Methods in Fluids. 1985;**5**:709-811

[21] Karamanev DG. Equation for calculation of the terminal velocity and drag coefficient of solid spheres and gas bubbles. Chemical Engineering Communications. 1996;**147**:73-79

[22] Dewsbury KH, Karamanev DG, Margaritis A. Rising solid hydrodynamics at high Reynolds numbers in non-newtonian fluids. Chemical Engineering Science. 2002;**87**: 120-133

[23] Ceylan K, Herdem S, Abbasov T. Theoretical model for estimation of drag forse in the flow non-newtonian fluids around spherical solid flow particles. Powder Technology. 1999;**103**: 286-295

[24] Dewsbury K, Karamanev DG, Margaritis SA. Hydrodynamic characteristics of free rise of light solid particles and gas bubbles in non-Newtonian liquids. Chemical Engineering Science. 1999;**54**:4825-4834

[25] Grace JR. Hydrodynamics of liquid drops in immiscible liquids. In: Cheremisinoff NP, Gupta E, editors. Handbook of Fluids in Motion. London: Ann Arbor Science; 1983. p. 273

[26] Yih C-S. Advances in Applied Mechanics. New York: Academic Press; 1972

[27] Karamanev DG. Equation for calculation of the terminal velocity and drag coefficient of solid spheres and gas bubbles. Chemical Engineering Communications. 1996;**147**:73-81

[28] Bubbles CRH. Drops and Particles in Non-Newtonian Fluids. Boca Raton: Taylor and Francis Group, CRC Press; 2006. p. 800

[29] Levich VG. Physico-Chemical Hydrodynamics. Moscow: Fizmatgiz; 1962

[30] Kelbaliev GI. Drag coefficient of variously shaped solid particles, drops and bubbles. Theoretical Foundations of Chemical Engineering. Springer. 2011; **45**(3):248-266

[31] Bharti RP, Chhabra RP, Eswaran V. Steady forced convection heat transfer from a heated circular cylinder to power-law fluids. International Journal of Heat and Mass Transfer. 2007;**50**:977-990

[32] Askovic R. Drag of a growing bubble at rectilinear accelerated ascension in pure liquids and binary solutions, Theoretical and Applied Mechanics. Belgrade. 2003;**30**(3):177-191

[33] Loth E. Particles, Drops and Bubbles: Fluid Dynamics and Numerical Methods. London: Cambridge University Press; 2010. p. 776

[34] Pinelli D, Magelli F. Solids falling and distribution in slurry reactors with dilute pseudoplastic suspension. Industrial and Engineering Chemistry Research. 2001;**40**:4456-4467

[35] Kelessids VC. An explicit equation for the terminal velocity of solid spheres falling in pseudoplastic liquids. Chemical Engineering Science. 2004;**59**: 4437-4446

[36] Astariata G, Marucci G, Nicolas I. Rheology Fluids. New York; 1980. p. 127

[37] Adina Ghirisan and Simion Draganm Sedimentation of concentrated suspensions in non-Newtonian fluids.// Universitatea Babes-Bolyai. 2011. http:// chem.ubbcluj.ro/~studiachemia/index.html

[38] Ovarlez G, Bertrand F, Coussot P. Xavier Chateau Shear-induced sedimentation in yield stress fluids. Journal of Non-Newtonian Fluid Mechanics, Elsevier. 2012;**177-178**:19-28

[39] Meyer RF. Prospects for heave grude oil development. Energy Exploration & Exploitation. 1987;**5**:27-85

[40] Nadyrov N. Problems of Processing Heavy Oils. Moscow: Chemistry; 1972

[41] Pedersen KS, Fredenslund A, Christensen PL, Thomassen P. Viscosity of crude oils. Chemical Engineering Science. 1984;**39**(6):1011-1016

[42] Soltani A, Nozarpour A, Aghamiri SF. Prediction of the kinematic viscosity of crude oil fraction. Petroleum Science and Technology. 2010;**28**:N6. DOI: 10.1080/10916460902804580

[43] Santos RG, Loh W, Bannwart AC, Trevisan OV. An overview of heavy oil properties and its recovery and transportation methods. Brazilian Journal of Chemical Engineering. 2014; **31**(03):571-590. DOI: 10.1590/0104-6632.20140313s00001853

[44] Alomair O, Jumaa M, Alkoriem A, Hamed M. Heavy oil viscosity and density prediction at normal and elevated Temperatures. Journal of Petroleum Exploration and Production Technologies. 2016v;**6**:253-263. DOI: 10.1007/s13202-015-0184-8

Chapter 3

The Effect of Ultrasonic Waves on Crude Oil Recovery

Ramin Tahmasebi-Boldaji

Abstract

In recent years, ultrasonic technology has played an important role in the development of oil fields, which has improved oil recovery. Ultrasonic waves are a very suitable method for producing oil at a low cost and without environmental pollution. The reservoir is treated using high-power sonication, which affects the physical properties of the oil and thus improves the permeability, which increases the oil recovery. The ultrasonic technique is also used to reduce the damage of the formation in the areas near the well, and this reduces the penetration of mud and sediments. However, ultrasonic waves remove oil barriers to the well and improve oil recovery for a long time. In this chapter, recent developments and laboratory and field results of ultrasonic waves in improving oil recovery will be discussed, and it will be shown that these waves are highly efficient.

Keywords: crude oil recovery, viscosity, ultrasonic waves, environmental pollution, ultrasound

1. Introduction

In the world, energy is the main condition for the development of human society. Oil, gas, and coal are the main sources of energy that have led to the development of human civilization in recent years [1]. Given that the demand for energy from these energy systems and the severe problems and challenges in the oil and gas industry are increasing day by day, the oil production capacity cannot meet this volume of demand. Also, because oil prices are high in the world, it has affected the oil market and the structure of international cooperation in this industry [1, 2]. However, the continuous reduction of hydrocarbons is an undeniable issue and serious efforts should be made to solve its problems [3]. Reduction in production can occur for two reasons: 1) reduction in reservoir pressure and 2) formation damage. Sedimentation of asphaltene and wax on rocks can occur due to the flow of drilling fluids, which is called formation damage. Damage formation is one of the main reasons for reducing the productivity of oil wells and has adverse effects on well production. Various techniques have been performed to eliminate formation damage. These techniques include high-pressure fractures and acid and solvent injections, which often have many disadvantages, including high cost, extensive facilities, and environmental problems [4, 5]. New and effective technologies have been developed to solve the mentioned problems in the oil industry. Ultrasonic oil recovery technology is one of the most effective solutions developed in the USA in the 1950s [6]. The ultrasonic technique and the use of its waves have desirable advantages such as low pollution, cheap, high efficiency, and environmentally friendly. Impurities and sediments in oil exploitation reduce oil flow and thus reduce production [7]. For this purpose, high-power ultrasonic waves are used to increase the oil permeability, which destroys the oil layer particles and increases the permeability by creating high sound acceleration [8]. Oil recovery technology means that the sonication operation leads to a change in the physical properties of the fluid, which improves the fluid flow status [1, 2] and reduces the pressure gradient of the porous medium [9].

In this chapter, the aim is to review recent developments in ultrasonic oil recovery technology, as well as the effect of these waves on the physical properties of the fluid and the viscosity of the oil.

2. Advantages of ultrasonic oil recovery technology

Due to the problems mentioned in the oil industry, the ultrasonic method is one of the suitable options for oil recovery. Oil recovery has the following advantages [7]:

- When using ultrasonic waves in areas of oil wells, it does not cause any pollution to the environment.
- The sonication operation is a simple and easy operation and has no complexity.
- The use of ultrasonic waves can be combined with other methods such as sonic water injection and acoustic acidification and increase the recovery of crude oil.
- The equipment used in crude oil recovery by ultrasonic waves is installed on vehicles and has a wide application, cheap, and high efficiency.

This equipment is shown in detail in **Figure 1**.

3. Effects of sonication operation

3.1 Mechanical vibration

Ultrasonic waves have a series of mechanical effects that change the velocity of elastic particles and cause agitation, loosening, scattering, and degassing. Mechanical vibrations also cause loss of cohesion between blocked particles, resizing capillary pores and reducing surface tension. These vibrations also cause very small cracks in the rocks of the formation and cause the separation of crude oil from the rocks. Impact pressure from high-frequency, high-power ultrasonic waves accelerates large molecules such as wax, asphaltene, and colloids, breaking the chain of molecules due to inertia [10–16].

3.2 Cavitation

The process of growth and collapse of hollow fluid bubbles due to sonication operations and changes in sound pressure is called cavitation. Pressures above 105 MPa can cause cavities to collapse [17]. This high pressure also causes secondary effects such as

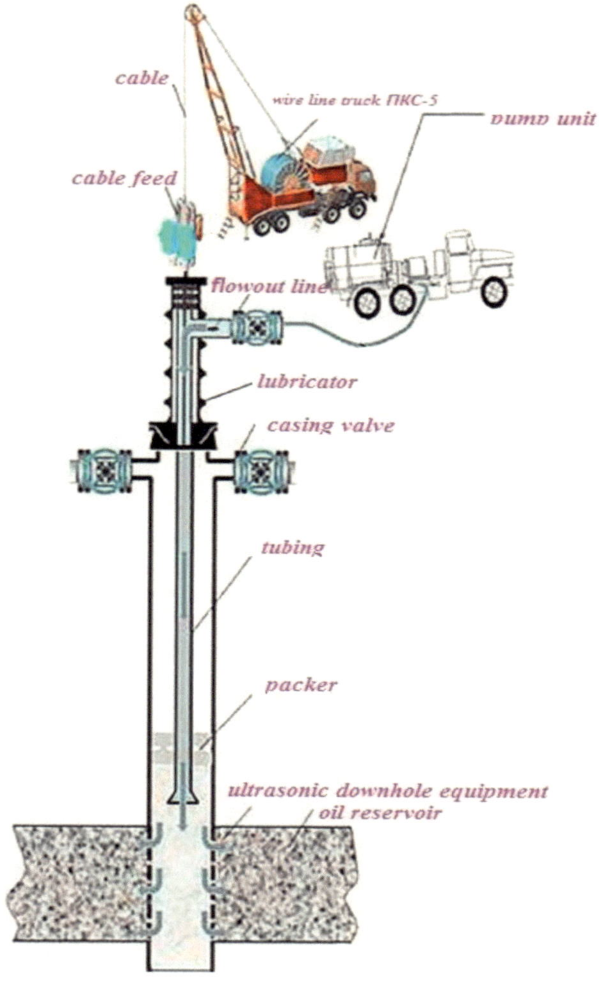

Figure 1.
The schematic of ultrasonic oil recovery device.

luminescence, phonation, ionization, and chemical reactions. One of the obstacles to oil flow is gas resistance. Gas resistance refers to a large number of gas cores. These gas nuclei combine to form larger bubbles, approaching the dredging target [18]. Frequent cavitation explosions occur in the fractures of the formation and on the solid surface, leading to high pressures and the consequent explosion of particles adhering to the surface. Also, due to the transverse flows of liquids and their alternating currents, particles are quickly removed from the surface [10]. However, cavitation breaks the molecular bonds of crude oil and reduces the molecular mass. Reducing the molecular mass reduces the viscosity and thus improves the fluidity of the crude oil.

3.3 Thermal action

The internal friction of crude oil is the reason for the resistance of the oil flow. As the viscosity of the oil decreases, the oil concentration also decreases, leading to improved flow. Therefore, to improve the flow of crude oil, its viscosity can be reduced, and this is possible by increasing the temperature. By absorbing ultrasonic waves, acoustic energy is converted into thermal energy, and also the boundary friction at the interface increases the temperature of the crude oil. However, at the moment of bubble collapse, a large amount of heat energy is released by cavitation, and with increasing ultrasonic frequency, the absorption of waves and boundary friction becomes more intense. Eventually, with increasing sonication power, cavitation and thermal energy become more significant [19–23].

4. Recent developments in sonication operations to improve oil recovery

In recent years, various devices have been invented that can improve the recovery of crude oil by ultrasonic waves. One type of environmentally friendly viscosity reduction is shown in **Figure 2** [24, 25]. This device has a very good anti-wax and antifouling feature and is used for underground oil pumping. This device can be used to increase the flow of crude oil, which increases pump efficiency and production. This device is also suitable for wells containing different wax and water and has good safety and no pollution.

One of the problems of the oil industry is the elimination of the scale of pipelines, and to solve this problem, an electromagnetic ultrasonic antifouling device was invented [26, 27]. The schematic of this device is shown in **Figure 3** [25].

Using an ultra-strong alternating magnetic field and ultrasonic sound field, the device affects the physical morphology and chemical properties of the crude oil sediments, dispersing and loosening the sediments completely and not adhering easily to the pipe wall. Also, with the invention of this device, they achieved many advantages, which are good antifouling effect, long life, no pollution, comfortable maintenance, and environmentally friendly [26].

Due to the fact that the viscosity reducing device cannot be used for more than 25% wax content, to solve this problem, a double sonic eddy current anti-wax viscosity reduction device was invented [28]. The schematic of this device is shown in **Figure 4** [25]. This device has useful capabilities, and a two-stage ultrasonic oscillator is used to improve the sound frequency and can destroy paraffin wax crystals and lead to a great reduction in viscosity. In addition, the use of eddy currents increases the fluidity of crude oil, prevents wax deposition, and increases oil flow capacity. In the oil industry, this device can be used for vertical, sloping, and horizontal oil production wells [28]. However, to improve the transmission efficiency of all oil production systems, three special types of cables are used, which are shown in **Figure 5** [25, 29].

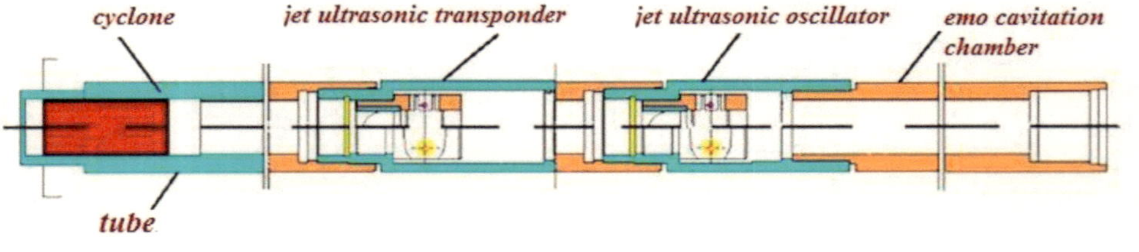

Figure 2.
The structure of a kind of environmentally friendly, anti-wax, and antiscaling viscosity reduction device [24, 25].

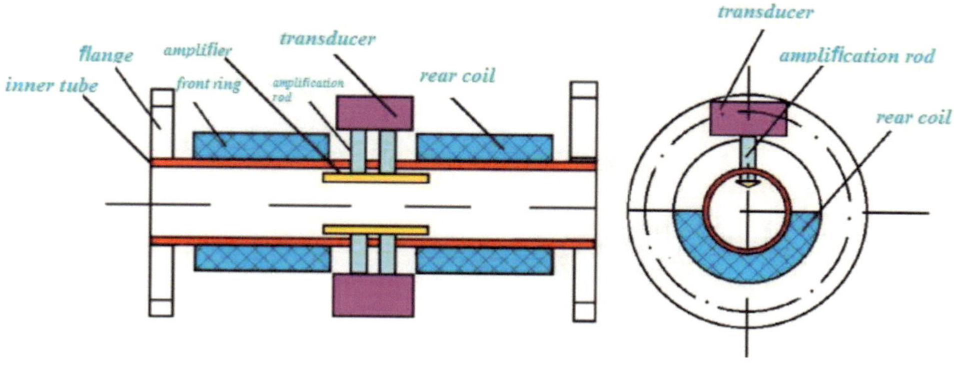

Figure 3.
The structure of electromagnetic ultrasonic antiscaling device [25].

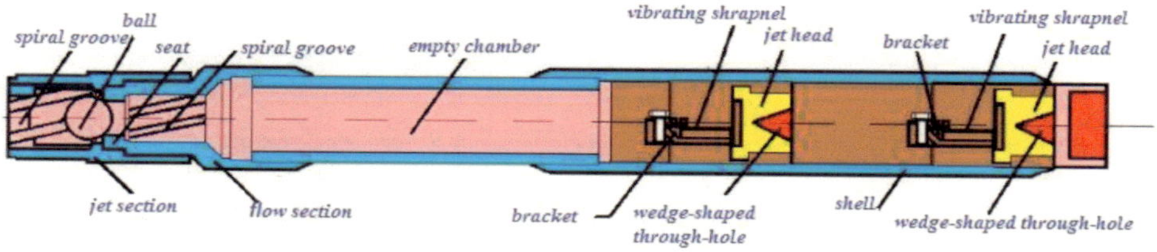

Figure 4.
The structure of double sonic eddy current anti-wax viscosity reduction device [25].

4.1 The effect of different ultrasonic parameters on viscosity and oil recovery efficiency

Recently, important studies and results have been obtained in the field of improving the recovery of crude oil using ultrasonic waves. Various ultrasonic parameters can have a significant impact on improving crude oil recovery. These parameters include cavitation, temperature, time, power, and frequency. Each of these parameters has a significant effect on crude oil recovery, which must be optimized for each specific operating condition of these parameters in order to achieve maximum recovery. In this section, the explanation and effect of these parameters are discussed.

4.1.1 The effect of ultrasonic temperature and cavitation

In a previous study, an ultrasound cleaning tank was used to separate oil from sludge [30]. Ultrasonic waves can separate solid particles from crude oil [31]. The results were obtained after deoiling the oil sludge, and the oil content increases when the temperature is above 40°C. However, too high temperature weakens cavitation and delays the separation of solid particles from crude oil [32]. That is, from the temperature range of 30 to 40°C, the cavitation intensity increases and reaches its maximum value at 40°C, and then with increasing temperature from 40 to 60°C, the cavitation intensity decreases and reaches a constant rate [30]. As a result, it can be said that cavitation has a maximum value at low temperature (40°C), and high temperature (>40°C) is suitable for molecular motion and increases the number of cavitation nuclei [30]. Also, the oil content increases in the range of acoustic pressure

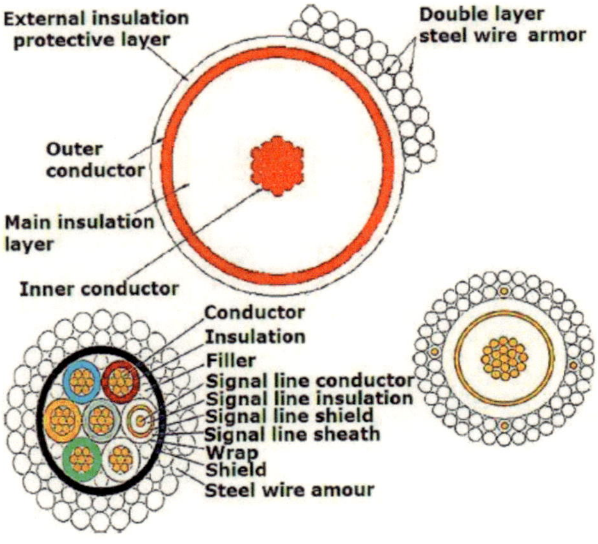

Figure 5.
Three kinds of special cables for improving transmission efficiency of the whole oil production system [25, 29].

above 0.10 MPa because cavitation increases with increasing acoustic pressure [33]. Temperature, power, and frequency parameters of 40° C, 0.1 MPa, and 25 kHz, respectively, can increase the oil recovery rate by 55.6% [30]. The optimal conditions for these parameters must be determined. The optimum conditions for oil sludge treatment for frequency, intensity, power, and soil-to-water ratio are 25 kHz, 0.33 W/Cm^2, 300 W, and 1/2, respectively [34]. With increasing temperature, the viscosity of the crude oil and the adhesion stress between the oil and the sand decrease. There is also a positive correlation between cavitation nuclei and sonication temperature. That is, the higher the temperature, the greater the number of cavitation nuclei. However, with a large increase in temperature, the internal pressure of the cavitation bubbles increases, which reduces the intensity of cavitation [35, 36].

With the radiation of ultrasonic waves, cavities are created in which large bubbles are created due to heat, and these waves cause these bubbles to collapse. The mechanical effect of this collapse causes the suspended conglomerates to disintegrate [37]. The increase in temperature and ultrasonic cavitation leads to the production of hydromechanical shear forces and causes disruption in macromolecules. These collapses that occur in the bubbles cause the creation of shear shock waves and solvent microjets that create turbulence in the surface layer around the solid particles and cause high local temperatures [38].

4.1.2 The effect of ultrasonic power

High ultrasonic power is one of the main features of ultrasound in crude oil recovery. Following recent research, an ultrasonic reactor was used to purify petroleum sludge [39]. The ultrasonic reactor was able to increase the oil recovery efficiency above 90% for some samples with a power of 240 W. Given that oil recovery increases with increasing power, the further increase of ultrasonic power not only does not improve oil recovery but also leads to reduced efficiency [39]. With a further increase in power, the ultrasonic recovery does not increase. It can be concluded that the cavitation phenomenon is responsible for the excretion of adsorbed molecules, and the effect of this phenomenon is affected by the size of the bubbles because larger bubbles can store more energy [35, 40, 41]. Ultrasonic cavitation is effective in inhibiting the growth of wax crystals. Ultrasonic cavitation causes long-chain alkanes to be converted to short-chain alkanes and C−C bonds are broken [42]. Also, after the irradiation of ultrasonic waves, the content of long-chain alkanes (C_{16}–C_{22}) decreases and short-chain alkanes (C_9 and C_{10}) are created [43]. With the increase of light hydrocarbons, wax solubility improves and the freezing point of crude oil decreases.

However, an optimal value for ultrasonic power must be determined, and increasing the power from a certain threshold does not help to improve oil recovery [35]. Very high levels of ultrasonic power can produce a lot of acoustic energy, and the energy is converted to smaller temperatures and pressures by small bubbles, and microscopic turbulence breaks the bonds and the adsorbed molecules are expelled [44]. Also, further increase of ultrasonic power due to inhibition of cavitation microbubble size cannot improve oil recovery [39]. However, ultrasonic power can be used to increase oil recovery, and the frequency parameter must be increased to increase recovery speed [39].

4.1.3 The effect of ultrasonic time

The combined method of ultrasound and thermochemical cleaning treatment for oily sludge was also investigated in another previous study [35]. After the operation, oil layers were collected for analysis. After 15 minutes of testing, the oil recovery is 99.28% and remains constant with a further increase in recovery time. Initially, under ultrasonic radiation, oil is desorbed from the surface of the sand and an oil–water emulsion is formed at a low concentration. At the beginning of the operation, the desorption rate is high, which leads to an increase in oil recovery. However, as the oil–water emulsion concentration increases, the oil re-adsorption on the removed sand surfaces increases at the same time. Also, when the desorption rate is equal to the re-adsorption, a further increase in sonication time does not increase oil recovery [35, 45]. As the ultrasonic radiation time increases, the bubbles inside the crude oil reach a critical size and then collapse. As the bubbles increase in size and collapse, the volume of crude oil increases, leading to a decrease in viscosity [37].

4.1.4 The effect of ultrasonic frequency

Another feature of ultrasound to increase crude oil recovery is low frequency (10–35 kHz). Reducing the frequency both increases cavitation and decreases the attenuation of acoustic energy [20]. However, the lower the frequency, the easier cavitation will occur [35]. Also, two types of 28 and 40 kHz generators were examined, and according to the results, ultrasound with a frequency of 28 kHz does better than 48 kHz and this is because [30]:

1. The 28-kHz frequency performs better in washing solid particles and the cavitation threshold at the 28-kHz frequency is less than 48 kHz, so the cavitation intensity is higher at 28 kHz and therefore the separation efficiency is higher at 28 kHz.

2. Koyusov research has shown that the optimal frequency for coagulation of solid particles is 21–25 kHz, and with increasing ultrasound frequency, the number of cavitations increases and causes strong vibrations.

Also, sonication operation has a significant effect on the viscosity of crude oil, which increases the fluidity, flow, and recovery of oil. In a study on a type of crude oil with a viscosity of 1250 MPa.s, wide ranges of frequencies from 18 to 25 kHz and power from 100 to 1000 W were investigated [46]. Ultrasonic frequencies of 18, 20, and 25 kHz can reduce the viscosity of oil by 480, 890, and 920 MPa.s, respectively [46]. Therefore, cavitation created by ultrasonic waves is able to break down heavy oil molecules into lighter hydrocarbons, and the power and irradiation time of the waves are the main parameters for reducing viscosity [46]. Another sample of crude oil with n-alkanes and tar-asphaltene compounds was examined at a resonant frequency of 24.3 kHz and a generator power of 4000 W [47]. Results were reported on the effect of ultrasonic waves on the viscosity of paraffin oils, and it was shown that these waves lead to a decrease in viscosity and pour point [47]. These results are also consistent with reports in other reports [48, 49] of reduced viscosity due to ultrasonic radiation. In another study, the rheological behavior of crude oil irradiated with ultrasonic waves was investigated [50]. Dissolution of heavy compounds in crude oil can be achieved by irradiation of ultrasonic waves with a frequency of 45 kHz and an optimal time of 45 minutes. In a recent study, we also found that the three parameters of time, power, and ultrasonic frequency have a significant effect on oil viscosity [51]. A sample of crude oil in oil reservoirs, named as Bangestan, at the Marun field, located in the south of Iran was employed for experiments. Brookfield dvz-lll ultra Rheometer was used to measure viscosity. The schematic of the equipment used is shown in **Figure 6** [51]. To produce ultrasonic waves of bath type, two ultrasonic generators were used, the first generator (YAXUN YX2000) with a frequency of 42 kHz and power of 35 and 50 W and the second generator (QUIGG SR 2014.16) with a frequency of 46 kHz and output power of 50 W. Simultaneous thermogravimetric and differential thermal analyzer (TGA/DTA Netzssh STA 409 PC LUXX) have been used to investigate the effect of ultrasonic waves on crude oil mass and thermal behavior. Crude oil viscosity was irradiated with ultrasonic waves at different times, frequencies, and power of sonication in ambient conditions, and its viscosity was measured after cooling of crude oil [51, 52]. As shown in **Figure 7** [51], the effects of three parameters of time, frequency, and ultrasonic power on the viscosity of crude

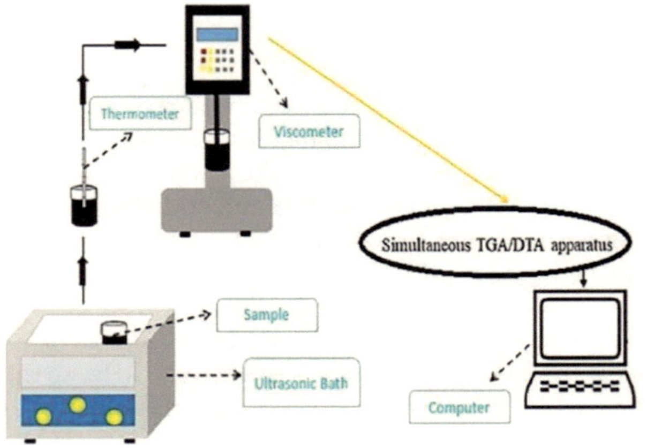

Figure 6.
Schematic of equipment used for investigating the effects of ultrasonic waves on the viscosity and thermal properties of crude oil [51].

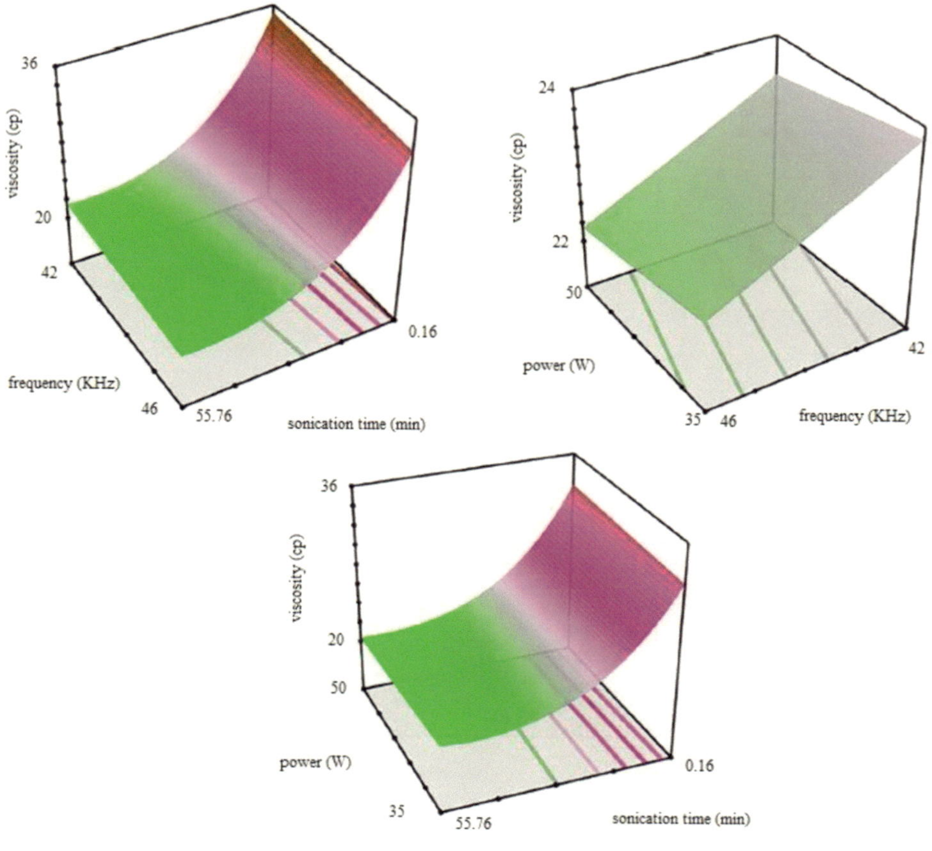

Figure 7.
The effect of triple ultrasonic factors on the viscosity of crude oil [51].

oil are significant. The sonication time has the greatest effect on the viscosity and with increasing the radiation time, the viscosity decreases sharply. On the other hand, the other two parameters have little effect on the viscosity of crude oil [51]. Good results in this field can be achieved by using appropriate statistical methods, modeling and artificial intelligence [53–59].

4.2 Mechanism of action of ultrasonic waves

In this section, the changes of molecular bonds of heavy crude oil due to ultrasonic radiation are analyzed. An example of a Fourier transform infrared spectroscopy (FTIR) spectrum of a sample of heavy crude oil is given in this section to investigate the effect of ultrasonic waves on the molecular bonds of crude oil (**Figure 8**) [60]. As can be seen in **Figure 8** [60], the angular vibrations of methyl and methylene are present in the wavelengths of 1378 and 1458 cm^{-1}, respectively. A carbon–carbon double bond is observed at wavelength of 1599 cm^{-1} and the tensile vibration of methylene at about 2854 cm^{-1} and for methyl at 2952 cm^{-1}. However, when heavy oil is exposed to ultrasonic radiation for 6 minutes, the peak intensity increases, indicating an increase in methyl functional groups. Also, the chains of heavy compounds are broken, leading to a decrease in viscosity. But, an increase in

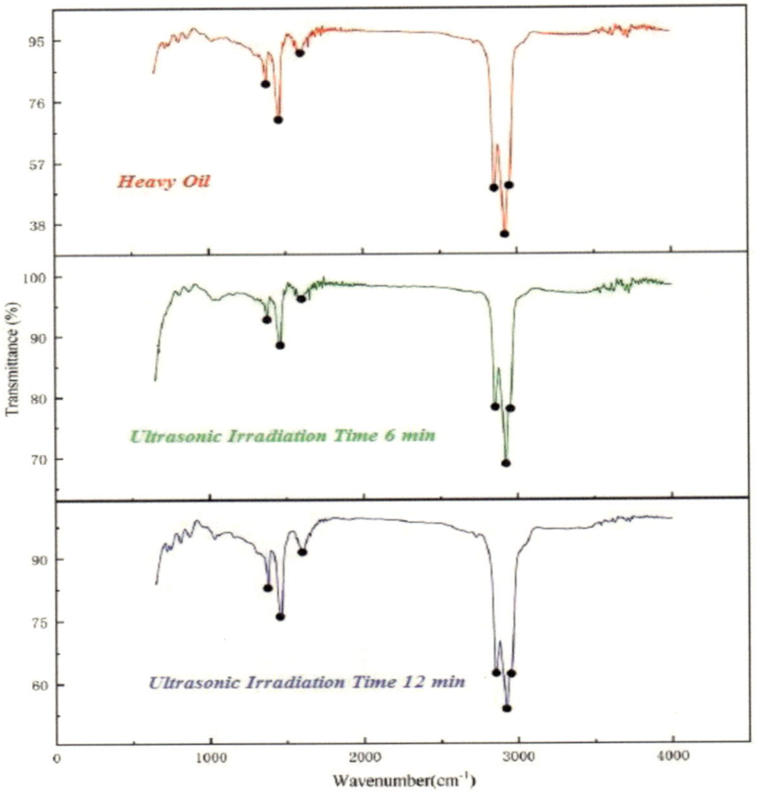

Figure 8.
FTIR spectra [60].

the time of ultrasonic radiation (12 min) reduces the intensity of the peaks. This means that due to the longer irradiation time, more long-chain molecules are converted to short-chain molecules, but due to excessive temperature and crossing the temperature threshold, light compounds evaporate and ultimately increase viscosity.

Severe cavitation can occur due to the intensity of ultrasonic radiation in heavy oil. In general, it can be said that acoustic cavitation has three main effects: mechanical, chemical, and thermal. Each of these three factors has a specific effect on oil. Thus, the mechanical agent creates a strong stirring effect and the thermal agent creates a high-temperature and high-pressure effect in the fluid. As mentioned, the contents of the functional groups increase as a result of sonication operations, which indicates that the molecular chains are broken and the heavy components of the oil are decomposed into lighter components. Reduction of heavy components and breaking of long chains due to wave radiation lead to a decrease in viscosity, which can be due to acoustic cavitation. Therefore, to better understand this issue, the carbon number distribution of three samples of heavy oil is given in **Figure 9** [60]. As the radiation time increases, the total amount of carbon, called long chains, gradually decreases. As shown in **Figure 9** [60], there are carbon chains of C_{40} and above C_{20} in heavy oil. As the ultrasonic radiation time increases, these chains are broken and the amount of carbon is reduced. This trend occurs in two other heavy oil samples, but in C_{11}–C_{20} carbon the number of light carbon chains decreases with increasing time. This is for

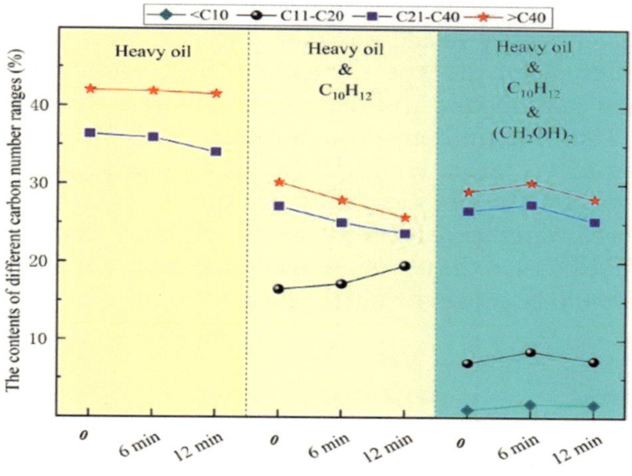

Figure 9.
Results of carbon number distribution [60].

the same reason as mentioned in the previous section, that as time increases, they produce excessive temperature waves that cause lighter compounds to evaporate and viscosity to increase.

5. Future views

Considering many studies have been performed on the effect of different ultrasonic parameters on increasing crude oil recovery, the ultrasonic cycle parameter has not been studied in all studies. It seems that different ultrasonic cycles can help a lot in improving the recovery of crude oil in the oil industry, in future laboratory and field studies. The ultrasonic cycle section is divided into active and passive intervals. In an ultrasonic homogenizer, the number of cycles is divided from 1 to 10. Each cycle represents a unique amount of active and passive intervals. In all recent studies, the effect of these active and passive intervals has not been investigated, which is a very important parameter in the recovery of crude oil.

6. Summary

This chapter examines recent developments and the effects of ultrasonic waves on viscosity and crude oil recovery. The results show that these waves with three mechanical, chemical, and thermal factors can improve oil recovery and break down heavy compounds and turn them into lighter compounds [10–23]. It is important to note that the various ultrasonic parameters (frequency, power, time, cycle, etc.) must be optimized. An optimal value can achieve the desired result. Ultrasonic waves also have great advantages such as cheapness, convenience, and environmental friendliness. It has also been concluded that sonication operations can decompose heavy colloidal compounds of crude oil and break down asphaltene molecules. Given that the presence of asphaltene in oil reservoirs can be problematic and clog pores and increase viscosity, it is important to know information about ultrasound parameters because these waves can be very useful in breaking down heavy molecules and reducing viscosity. In addition to the effect of the mentioned parameters on the recovery of crude oil, the type of these waves is also important. This means that intermittent or continuous use of these waves can have different effects on crude oil recovery [61]. Thus, intermittent sonication operations lead to greater crude oil recovery than continuous sonication [61]. In addition to increasing the permeability of ultrasonic waves [62], the vibrations of these waves can also affect the flow of oil and lead to improved water drive recovery and oil structure changes and viscosity decreases [63, 64]. However, the use of another parameter, the ultrasonic cycle, can be effective and should be researched in the future.

Conflict of interest

The authors declare no conflict of interest.

Author details

Ramin Tahmasebi-Boldaji
Department of Chemical Engineering, College of Engineering, University of Isfahan, Isfahan, Iran

*Address all correspondence to: ramintahmasbi68@gmail.com

References

[1] Alhomadhi E, Amro M, Almobarky M. Experimental application of ultrasound waves to improved oil recovery during waterflooding. Journal of King Saud University-Engineering Sciences. 2014;**26**(1):103-110

[2] Zhu T, Huang X, Vajjha PK. Downhole harmonic vibration oil-displacement system: a new IOR tool. In: SPE Western Regional Meeting. OnePetro; 2005

[3] Hamidi H, Rafati R, Junin RB, Manan MA. A role of ultrasonic frequency and power on oil mobilization in underground petroleum reservoirs. Journal of Petroleum Exploration and Production Technology. 2012;**2**(1):29-36

[4] Hays J, Finkel ML, Depledge M, Law A, Shonkoff SB. Considerations for the development of shale gas in the United Kingdom. Science of the Total Environment. 2015;**512**:36-42

[5] Werner AK, Vink S, Watt K, Jagals P. Environmental health impacts of unconventional natural gas development: A review of the current strength of evidence. Science of the Total Environment. 2015;**505**:1127-1141

[6] Abdulrahman MM, Meribout M. Antenna array design for enhanced oil recovery under oil reservoir constraints with experimental validation. Energy. 2014;**66**:868-880

[7] Mullakaev MS, Abramov VO, Abramova AV. Ultrasonic automated oil well complex and technology for enhancing marginal well productivity and heavy oil recovery. Journal of Petroleum Science and Engineering. 2017;**159**:1-7

[8] Wang Z, Xu Y, Gu Y. A light lithium niobate transducer design and ultrasonic de-icing research for aircraft wing. Energy. 2015;**87**:173-181

[9] Jia Z. Research on the Technology of Paraffin-Controlling and Viscosity-Reducing for Crude Oil Based on the Utransonic Transducer Array. Harbin Institute of Technology; 2009

[10] Xianyong D, Ping Z. Study on viscosity reduction test in Shengli offshore oil ultrasonic. Oil Gas Storage Transportation. 2004;**23**(3):32-35

[11] Jihu W, Guokun Q, Longtang Z. Study on the effect of ultrasound on the oil viscosity. Journal of Shengli Oilfield Staff University. 2006:1008-8083

[12] Mohsin M, Meribout M. An extended model for ultrasonic-based enhanced oil recovery with experimental validation. Ultrasonics Sonochemistry. 2015;**23**:413-423

[13] Hamidi H, Mohammadian E, Junin R, Rafati R, Azdarpour A, Junid M, et al. The effect of ultrasonic waves on oil viscosity. Petroleum Science and Technology. 2014;**32**(19):2387-2395

[14] Hamidi H, Mohammadian E, Junin R, Rafati R, Manan M, Azdarpour A, et al. A technique for evaluating the oil/heavy-oil viscosity changes under ultrasound in a simulated porous medium. Ultrasonics. 2014;**54**(2):655-662

[15] Wang Z, Xu Y, Bajracharya S. The comparison of removing plug by ultrasonic wave, chemical deplugging agent and ultrasound–chemical combination deplugging for near-well ultrasonic processing technology. Ultrasonics Sonochemistry. 2015;**27**:339-344

[16] Wang Z, Gu S, Zhou L. Research on the static experiment of super heavy crude oil demulsification and dehydration using ultrasonic wave and audible sound wave at high temperatures. Ultrasonics Sonochemistry. 2018;**40**:1014-1020

[17] Pu C, Shi D, Zhao S, Xu H. Technology of removing near wellbore inorganic scale damage by high power ultrasonic treatment. Petroleum Exploration and Development. 2011;**38**(2):243-248

[18] Bin L, Jiteng G, Jianguo Z. Sound application of physics in crude oil paraffin control and viscosity reduction of the device. China Petrolium Chemical Industry. 2002;**31**(8):517-520

[19] Hongxing X. Experimental Study on Ultrasonic Treatment for Rremoval of Near Wellbore Damage and Technological Parameters Optimization. East China: College of Petroleum Engineering China University of Petroleum; 2009

[20] Wang Z, Zeng J, Song H, Li F. Research on ultrasonic excitation for the removal of drilling fluid plug, paraffin deposition plug, polymer plug and inorganic scale plug for near-well ultrasonic processing technology. Ultrasonics Sonochemistry. 2017;**36**:162-167

[21] Kotyusov AN, Nemtsov BE. Induced coagulation of small particles under the action of sound. Acta Acustica United with Acustica. 1996;**82**(3):459-463

[22] Kou J. Research on heavy oil dehydration of ultrasonic. Oil Gas Fields Engineering. 2009;**8**:273-281

[23] Antes FG, Diehl LO, Pereira JS, Guimarães RC, Guarnieri RA, Ferreira BM, et al. Feasibility of low frequency ultrasound for water removal from crude oil emulsions. Ultrasonics Sonochemistry. 2015;**25**:70-75

[24] Wang B, Wang C. A kind of environmentally-friendly, anti-wax and antiscaling viscosity reduction device, CN 101328796A. 2008

[25] Wang Z, Fang R, Guo H. Advances in ultrasonic production units for enhanced oil recovery in China. Ultrasonics Sonochemistry. 2020;**60**:104791

[26] Zhang Z. Electromagnetic ultrasonic anti-scaling device. CN 101376132A. 2009

[27] Abramov VO, Abramova AV, Bayazitov VM, Mullakaev MS, Marnosov AV, Ildiyakov AV. Acoustic and sonochemical methods for altering the viscosity of oil during recovery and pipeline transportation. Ultrasonics Sonochemistry. 2017;**35**:389-396

[28] Zhou P, Xu Y. Double sonic eddy current anti-wax viscosity reduction device. CN 203769723 U. 2014

[29] Kawamata A, Hosaka H, Morita T. Non-hysteresis and perfect linear piezoelectric performance of a multilayered lithium niobate actuator. Sensors and Actuators A: Physical. 2007;**135**(2):782-786

[30] Xu N, Wang W, Han P, Lu X. Effects of ultrasound on oily sludge deoiling. Journal of Hazardous Materials. 2009;**171**(1-3):914-917

[31] Bougrier C, Carrère H, Delgenes JP. Solubilisation of waste-activated sludge by ultrasonic treatment. Chemical Engineering Journal. 2005;**106**(2):163-169

[32] Canoğlu S, Gültekin BC, Yükseloğlu SM. Effect of ultrasonic

energy in washing of medical surgery gowns. Ultrasonics. 2004;**42**(1-9):113-119

[33] Behrend O, Schubert H. Influence of hydrostatic pressure and gas content on continuous ultrasound emulsification. Ultrasonics Sonochemistry. 2001;**8**(3):271-276

[34] Gao Y, Ding R, Wu S, Wu Y, Zhang Y, Yang M. Influence of ultrasonic waves on the removal of different oil components from oily sludge. Environmental Technology. 2015;**36**(14):1771-1775

[35] Jin Y, Zheng X, Chu X, Chi Y, Yan J, Cen K. Oil recovery from oil sludge through combined ultrasound and thermochemical cleaning treatment. Industrial & Engineering Chemistry Research. 2012;**51**(27):9213-9217

[36] Ji-Sheng YANG, Hui X. Experimental research on crude oil removal from oily sludge and sands by ultrasound. Acta Petrolei Sinica (Petroleum Processing Section). 2010;**26**(2):300

[37] Taheri-Shakib J, Shekarifard A, Naderi H. The experimental investigation of effect of microwave and ultrasonic waves on the key characteristics of heavy crude oil. Journal of Analytical and Applied Pyrolysis. 2017;**128**:92-101

[38] Harrison ST. Bacterial cell disruption: A key unit operation in the recovery of intracellular products. Biotechnology Advances. 1991;**9**(2):217-240

[39] He S, Tan X, Hu X, Gao Y. Effect of ultrasound on oil recovery from crude oil containing sludge. Environmental Technology. 2019;**40**(11):1401-1407

[40] Breitbach M, Bathen D, Schmidt-Traub H. Effect of ultrasound on adsorption and desorption processes. Industrial & Engineering Chemistry Research. 2003;**42**(22):5635-5646

[41] Li XB, Liu JT, Xiao YQ, Xiao X. Modification technology for separation of oily sludge. Journal of Central South University. 2011;**18**(2):367-373

[42] Wang Z, Gu S. State-of-the-art on the development of ultrasonic equipment and key problems of ultrasonic oil prudction technique for EOR in China. Renewable and Sustainable Energy Reviews. 2018;**82**:2401-2407

[43] Zhang J. Research on the technology of paraffin-controlling and viscosity-reducing for crude oil based on the ultrasonic transducer array. 2009

[44] Pilli S, Bhunia P, Yan S, LeBlanc RJ, Tyagi RD, Surampalli RY. Ultrasonic pretreatment of sludge: A review. Ultrasonics Sonochemistry. 2011;**18**(1):1-18

[45] Feng D, Aldrich C. Sonochemical treatment of simulated soil contaminated with diesel. Advances in Environmental Research. 2000;**4**(2):103-112

[46] Wang Z, Xu Y, Gu Y. Lithium niobate ultrasonic transducer design for enhanced oil recovery. Ultrasonics Sonochemistry. 2015;**27**:171-177

[47] Mullakaev MS, Volkova GI, Gradov OM. Effect of ultrasound on the viscosity-temperature properties of crude oils of various compositions. Theoretical Foundations of Chemical Engineering. 2015;**49**(3):287-296

[48] Aliev F, Mukhamatdinov I, Kemalov A. The influences of ultrasound waves on rheological and physico-chemical properties of extra heavy oil from "Ashalcha" field. International Multidisciplinary Scientific

GeoConference: SGEM. 2017;**17**(1.4):941-948

[49] Rahimi MA, Ramazani SAA, Alijani Alijanvand H, Ghazanfari MH, Ghanavati M. Effect of ultrasonic irradiation treatment on rheological behaviour of extra heavy crude oil: A solution method for transportation improvement. The Canadian Journal of Chemical Engineering. 2017;**95**(1):83-91

[50] Mousavi SM, Ramazani A, Najafi I, Davachi SM. Effect of ultrasonic irradiation on rheological properties of asphaltenic crude oils. Petroleum Science. 2012;**9**(1):82-88

[51] Tahmasebi Boldaji R, Rajabi Kuyakhi H, Tahmasebi Boldaji N, Rajabzadeh M, Rashidi S, Torki M, et al. A comparative study of mathematical and ANFIS models to determine the effect of ultrasonic waves on the viscosity of crude oil. Petroleum Science and Technology. 2022;**40**(2):150-165

[52] Razavifar M, Qajar J. Experimental investigation of the ultrasonic wave effects on the viscosity and thermal behaviour of an asphaltenic crude oil. Chemical Engineering and Processing-Process Intensification. 2020;**153**:107964

[53] Tahmasebi-Boldaji R, Hatamipour MS, Khanahmadi M, Sadeh P, Najafipour I. Ultrasound-assisted packed-bed extraction of hypericin from Hypericum perforatum L. and optimization by response surface methodology. Ultrasonics Sonochemistry. 2019;**57**:89-97

[54] Tahmasebi Boldaji R, Rajabi Kuyakhi H, Tahmasebi Boldaji N. Prediction of 1-butanol and diesel fuel blend heat capacity by response surface methodology. Petroleum Science and Technology. 2020;**38**(11):737-743

[55] Boldaji RT, Kuyakhi HR. Predicting supercritical extraction of St. John's wort by simple quadratic polynomial model and adaptive neuro-fuzzy inference system-firefly algorithm. Journal of Analytical Techniques and Research. 2021;**3**(2):14-27

[56] Rajabi Kuyakhi H, Boldaji RT, Azadian M. Light hydrocarbons solvents solubility modeling in bitumen using learning approaches. Petroleum Science and Technology. 2021;**39**(4):115-131

[57] Kuyakhi HR, Zarenia O, Boldaji RT. Hybrid intelligence methods for modeling the diffusivity of light hydrocarbons in bitumen. Heliyon. 2020;**6**(9):e04936

[58] Rajabi Kuyakhi H, Tahmasebi Boldaji R. Developing an adaptive neuro-fuzzy inference system based on particle swarm optimization model for forecasting Cr (VI) removal by NiO nanoparticles. Environmental Progress & Sustainable Energy. 2021;**40**(4):e13597

[59] Rajabi Kuyakhi H, Tahmasbi Boldaji R. A novel ANFIS model to prediction of the density of n-alkane in different operational condition. Petroleum Science and Technology. 2019;**37**(24):2429-2434

[60] Gao J, Li C, Xu D, Wu P, Lin W, Wang X. The mechanism of ultrasonic irradiation effect on viscosity variations of heavy crude oil. Ultrasonics Sonochemistry. 2021;**81**:105842

[61] Agi A, Junin R, Chong AS. Intermittent ultrasonic wave to improve oil recovery. Journal of Petroleum Science and Engineering. 2018;**166**:577-591

[62] Agi A, Junin R, Shirazi R, Afeez G, Yekeen N. Comparative study of ultrasound assisted water and

surfactant flooding. Journal of King Saud University-Engineering Sciences. 2019;**31**(3):296-303

[63] Mullakaev MS, Abramov VO, Abramova AV. Development of ultrasonic equipment and technology for well stimulation and enhanced oil recovery. Journal of Petroleum Science and Engineering. 2015;**125**:201-208

[64] Wang Z, Xu Y. Review on application of the recent new high-power ultrasonic transducers in enhanced oil recovery field in China. Energy. 2015;**89**:259-267

Chapter 4

Water Content of Oil-Water Mixtures by the Speed of Sound Measurement

Alcir de Faro Orlando

Abstract

Ultrasonic meters are calibrated for flow rate measurement of pure oil flows. However, the indicated oil flow rate is greater than its true value when water is mixed with oil in the flow, as usual in pipeline flows. A methodology was developed to continuously measure the water content of oil–water mixtures, using the already measured speed of sound by installed intrusive or non-intrusive (clamp on type) ultrasonic meters for flow rate measurement, together with the previously determined speed of sound and density of pure water and oil as a function of temperature, thus avoiding the traditional method of laboratory analysis of samples. This chapter shows a reduction of the uncertainty of measurement when the meter indicated speed of sound in the oil–water mixture is directly correlated to the traceable measurements of water content and temperature of previously prepared oil–water mixtures, what turns out to be the meter calibration for water content measurement. Water content uncertainty value of 0.0025 is obtained when fitting a curve to the experimental data by the least square method in the 0–0.025 range that covers the fiscal measurement range (<0.01).

Keywords: water content, oil–water mixtures, speed of sound, fiscal flow rate measurement, ultrasonic flow measurement

1. Introduction

The most common method of measuring the amount of water in an oil–water mixture in commercial use is the measurement of the dielectric constant by a capacitance probe. However, capacitance probes are subject to coating by paraffin and other substances, which render them inaccurate in a short period of time. Another scheme involves the laboratory analysis of samples, but this method is labor intensive, requires much time to complete, and does not lend itself to continuous monitoring of fluid, such as crude oil flowing in a pipeline [1].

Oil is customarily metered as it flows from a producing line to a customer, either by a pipeline company or a refiner. Since oil is an expensive commodity, it is important that it is accurately measured, and if it is mixed with water, the customer pays for the water. If the ratio of oil and water is known, the water content can be deducted [1].

When installed in pipelines, some ultrasonic meters of the spool type can accurately measure the liquid flow rate to within 0.2%, as required for fiscal measurement. Furthermore, they also measure the speed of sound in the flowing liquid, which can be used to estimate its water content. Ultrasonic meters of the clamp on type, with two acoustic paths, are less accurate for flow rate measurement. However, they can accurately measure the speed of sound in any position along the pipeline, as a non-intrusive meter, what makes then convenient for inspecting the fluid flow. They are calibrated for flow rate measurement of pure oil flows. However, the indicated oil flow rate is larger than its true value when water is mixed with oil in the flow. A maximum of 1% water content of the fluid, for example, is required for fiscal measurement operations. They indicate at the same time both flow rate and speed of sound values, thus making easy to estimate the pure oil flow rate from water content measurement.

The US patents [1–3] show that the water content of the oil–water mixture is a function of the speed of sound. In this chapter, an expression was theoretically developed, using the previously measured speed of sound and density values for pure water and pure oil, for calculating the water content of the oil–water mixture, without sampling the fluid and taking it to a laboratory for analysis. The methodology was experimentally verified with the measured speed of sound by two types of flow meters, namely (a) Eight (8) acoustic path intrusive ultrasonic flow meter and (b) Two (2) acoustic path non-intrusive clamp on type ultrasonic flow meter. A methodology was also developed to correlate directly the water content to the measured fluid speed of sound for several previously prepared diesel fuel-water mixtures, thus reducing the propagation of the uncertainty of measurement from pure fluid measured properties. The methodology is detailed in Ref. [4].

2. Measurement of fluid properties

2.1 Water content

An expression has been developed in this chapter to relate directly the oil–water mixture water content to the product of density (ρ) and speed of sound (c), measured for the mixture and previously determined for pure oil and pure water, as a function of temperature and pressure.

The water content of an oil–water mixture (f) is defined as the ratio between the mass of water (m_{water}) and the total mass of the mixture (m), which can be calculated as the sum of the mass of water (m_{water}) and the mass of oil (m_{oil}), Eq. (1). The specific volume of either water (v_{water}) or oil (v_{oil}) can be calculated as the ratio between the volume (V) and the mass (m), respectively, for water ($v_{\text{water}} = V_{\text{water}}/m_{\text{water}}$) and oil ($v_{\text{oil}} = V_{\text{oil}}/m_{\text{oil}}$). Density is calculated as the inverse of the specific volume, for either water (ρ_{water}) or oil (ρ_{oil}). Thus, the following equations can be written for the water content (f) and specific volume of the mixture (v)

$$f = \frac{m_{water}}{m} = \frac{m_{water}}{m_{water} + m_{oil}} \tag{1}$$

$$v = v_{oil} - f\left(v_{oil} - v_{water}\right) \tag{2}$$

The speed of sound in a fluid (c) can be defined in terms of the partial derivative of the pressure (P) with respect to its specific volume (v), at constant entropy (s)

$$c = \sqrt{\left.\frac{\partial P}{\partial \rho}\right|_s} = \frac{1}{\rho}\sqrt{-\left.\frac{\partial P}{\partial v}\right|_s} \tag{3}$$

From Maxwell relation and using Eq. (3),

$$\left.\frac{\partial v}{\partial P}\right|_s = 1/\left.\frac{\partial P}{\partial v}\right|_s = -1/(\rho\, c)^2 \tag{4}$$

Substituting v, ρ, and c for, respectively, v_{water}, ρ_{water}, and c_{water}, Eq. (4) is valid for water. Likewise, substituting v, ρ, and c for, respectively, v_{oil}, ρ_{oil}, and c_{oil}, Eq. (4) is valid for oil.

Taking the partial derivative of Eq. (2) with respect to P, at constant entropy (s),

$$f = \frac{\left[\frac{1}{(\rho.c)^2_{oil}} - \frac{1}{(\rho.c)^2}\right]}{\left[\frac{1}{(\rho.c)^2_{oil}} - \frac{1}{(\rho.c)^2_{water}}\right]} \tag{5}$$

Eq. (5) shows that the water content (f) of the fluid mixture can be calculated from the measured speed of sound (c) if both density and speed of sound have been determined for pure water and pure oil before measurement. The mixture density can be calculated by substituting, in Eq. (2), the specific volumes of the mixture (v), pure oil (v_{oil}), and pure water (v_{water}) for their inverse, which are, respectively, the densities of the mixture (ρ), pure oil (ρ_{oil}), and pure water (ρ_{water}), resulting in Eq. (6),

$$\rho = \frac{\rho_{oil}}{1 - f\left(1 - \frac{\rho_{oil}}{\rho_{water}}\right)} \tag{6}$$

2.2 Water density

The water density (ρ_{water}), in kg/m^3, as a function of temperature (T), in °C, was modeled [5] by Eq. (7) in the 0–40°C range, with an uncertainty interval in the 0.00084 to 0.00088 kg/m^3 range, for a 95.45% level of confidence, using the coefficients of **Table 2**.

$$\rho_{water} = A_5.\left[1 - \frac{(T + A_1)^2.(T + A_2)}{A_3.(T + A_4)}\right] \tag{7}$$

2.3 Speed of sound in water

The speed of sound in distilled water (c_{water}), in m/s, was measured by NIST [6] at nearly atmospheric pressures, with an uncertainty (95.45%) of 0.05 m/s, and modeled as a function of temperature (T), in °C, by the Eq. (8) using the coefficients of **Table 1**.

$$c_{water} = \sum_{i=0}^{5} A_i\, T^i \tag{8}$$

The speed of sound (c_{water}) in salt water has been measured and correlated by different authors to temperature (0–40°C), salinity (0 to 4.2%), and pressure or water depth (0 to 1000 bar). The available equations are classified as (a) Simple, Mackenzie [7] and Coppens [8] and (b) Refined, UNESCO [9], Del Grosso [10] and NPL [11]. The speed of sound was calculated by those equations, reduced to the polynomial representation by Eq. (8) for zero salinity and atmospheric pressure, and compared to those obtained for distilled water, NIST [6] and for tap water, measured in this research with a root mean square dispersion of 1.58 m/s (95.45% level of confidence). Due to a restricted range of validity of some equations for salinity, only two equations were chosen for comparison, UNESCO [9] and NPL [11]. **Table 1** shows the coefficients to be used in Eq. (8).

Figure 1 shows the variation of the speed of sound in water with temperature for different equations, using in Eq. (8) the coefficients of **Table 1**. The 20–30°C temperature range was chosen because it represents the temperature conditions of the experiments. It can be seen that the NIST equation [6], for distilled water, calculates the largest values. The smallest values are calculated for tap water (this research), resulting in a systematic error in the 2.3 to 4.3 m/s range (0.16 to 0.29% range). The UNESCO equation [9] and NPL equation [11] calculate similar values, which lie in between the two first ones. Considering that the standard deviation of the speed of sound measurement during the experiments is in the 0.16 to 0.21 m/s, it can be concluded that the difference is not due to meter repeatability, but possibly to the fact that the tap water does not have the same properties of the distilled water, or there is a systematic error.

The speed of sound in water was measured by a two acoustic path non-intrusive type (clamp on) ultrasonic meter and by an eight acoustic path intrusive type (spool) ultrasonic meter. The measurement reliability was verified by varying the flow rate, the acoustic path inclination angle, and the velocity profile development. As expected, the difference was within the uncertainty of measurement, because of the fact that the speed of sound is a fluid property and is not influenced by the flow pattern.

During the tests for the clamp on type meter, the temperature varied in the (23.3 ± 1.1)°C range. The average speed of sound was 1487.3 m/s. According to the calculated values by Eq. (8) and **Table 1**, the speed of sound would be 1492.0 m/s for the UNESCO and NPL equations, 1490.3 m/s for tap water, and 1493.2 m/s for the NIST equation. All values were inside the interval about the calculated value for tap water (1490.3 ± 3.0) m/s.

During the tests for the spool type meter, the temperature varied in the (23.3 ± 1.5)°C range. The average speed of sound was 1490.8 m/s range. According to the calculated values by Eq. (8) and **Table 1**, the speed of sound would be 1492.0 m/s

Coefficients	UNESCO	NPL	Tap water	NIST
A_0	1402.388	1402.5	1407.479	1402.736
A_1	5.0383	5	4.33597	5.03358
A_2	−5.8109E-02	−5.44E-02	−3.467E-02	−5.79506E-02
A_3	3.3432E-04	2.1E-04		3.31636E-04
A_4	−1.47797E-06			−1.45262E-06
A_5	3.1419E-09			3.0449E-09

Table 1.
Coefficients for calculating the speed of sound in water by Eq. (8).

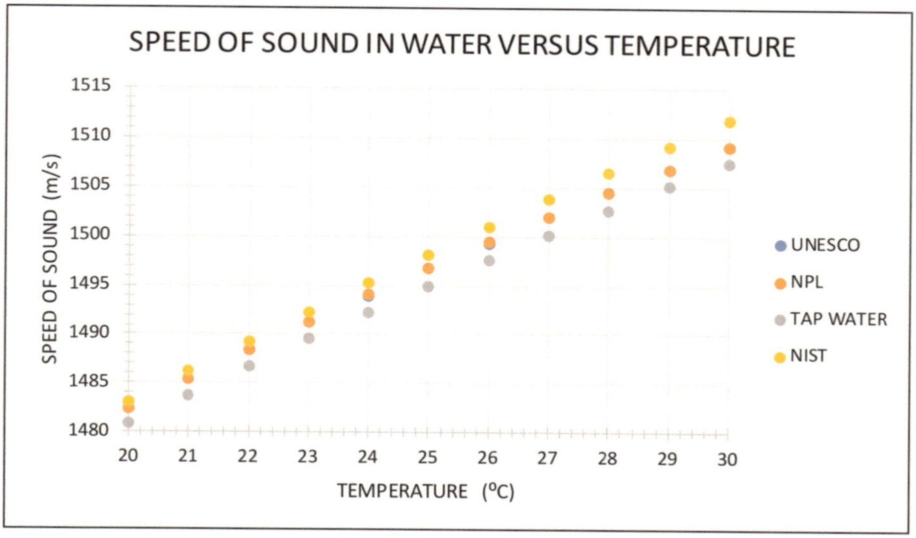

Figure 1.
Speed of sound in water for different equations in the 20–30°C range.

A_1	A_2	A_3	A_4	A_5
−3.983035	301.797	522528.9	69.34881	999.974950

Table 2.
Coefficients for calculating water density by Eq. (7).

for the UNESCO and NPL equations, 1490.3 m/s for tap water, and 1493.2 m/s for the NIST equation. All values were inside the interval about the calculated value for tap water (1490.3 ± 2.9) m/s.

Due to the fact that no information was available for the tap water composition used in the experiments of this research, the interval (1490.3 ± 3.0) m/s, about the mean measured value by the clamp on type meter, obtained by curve fitting the experimental data of this research with temperature, Eq. (8), is considered to be the upper limit of the uncertainty interval, that is, 0.20% of the measured speed of sound by the clamp on type ultrasonic meter (**Table 1**).

Oil type	ρ_{min}	ρ_{max}	K_o	K_1	K_2
	kg/m³	**kg/m³**	**°F⁻¹**	**°F⁻¹**	**°F⁻¹**
Crude	610.6	1163.5	341.0957	0.00000	0.0000
Fuel	838.3127	1163.5	103.8720	0.27010	0.0000
Jet Fuel	787.5195	838.3127	330.3010	0.00000	0.0000
Transition	770.3520	787.5195	1489.0670	0.00000	−0.0018684
Gasoline	610.6	770.3520	192.4571	0.24380	0.0000
Lubricating	800.9	1163.5	0.0000	0.34878	0.0000

Table 3.
Coefficients for calculating α_{60} by Eq. (9).

2.4 Oil density

The API methodology [12] calculates the oil density at temperature (T) and pressure (P) from a measured value at other conditions, to within 0.25%, up to 200°F (93.33°C). The following parameters are used in Eqs. (9)–(13), **Tables 3** and **4**.

- ρ_{oil} (kg/m^3) oil density
- T (°F) oil temperature
- P (psi) oil pressure
- α_{60} (°F^{-1}) oil expansion coefficient at 60°F
- ρ_{60} (kg/m^3) oil density at 60°F and atmospheric pressure
- CTL density correction from 60°F to T
- CPL density correction from atmospheric pressure to P

The API methodology was utilized to produce, iteratively, diesel fuel density values in the 5–35°C range, every 1°C, from a initially measured value of 856.12 kg/m^3 at 23.31°C, which was used to calculate α_{60}. The density values were curve fitted, resulting in Eq. (14) and **Table 5**, with a curve fitting error smaller than 0.001 kg/m^3.

$$\alpha_{60} = \frac{K_0}{\rho_{60}^2} + \frac{K_1}{\rho_{60}} + K_2 \tag{9}$$

$$CTL = EXP\{-\alpha_{60}.(T-60).[1+0.8.\alpha_{60}.(T-60)]\} \tag{10}$$

$$F_p = EXP\left\{A + B.T + \frac{C + D.T}{\rho_{60}^2}\right\} \tag{11}$$

$$CPL = \frac{1}{1 - F_p.(P - PE)/100000} \tag{12}$$

$$\rho_{oil} = CTL.CPL.\rho_{60} \tag{13}$$

A	B	C	D	PE
	°F^{-1}	(kg/m^3)^2	(kg/m^3)^2.°F^{-1}	psi
−19.947	0.00013427	793.920	2326	0

Table 4.
Coefficients for calculating F_p by Eq. (11).

A_0	A_1	A_2
872.481649	−0.697928	−0.000168

Table 5.
Coefficients for calculating oil density by Eq. (14).

$$\rho_{oil} = \sum_{i=0}^{2} A_i\, T^i \tag{14}$$

2.5 Speed of sound in oil

The speed of sound in oil can be calculated by the ARCO formula [13], which is a function of the oil density (kg/m^3), calculated by Eq. (15) in API units (**Table 6**),

$$API = \frac{141.5}{\rho_{oil}/999.016} - 131.5 \tag{15}$$

$$c = (K_a.6894.757/\rho_{oil})^{½} \tag{16}$$

$$K_a = A + B.P - C.(T + 459.67)^{½} - D.API - E.API^2 + F.(T + 459.67).API \tag{17}$$

where

- c (m/s) speed of sound
- ρ_{oil} (kg/m^3) oil density
- T (^{0}F) oil temperature
- P (psi) oil pressure
- $K_a(psi)$ adiabatic compressibility modulus.

Diesel fuel was used as fluid for the experiments in this research for being readily available and having its properties well studied. In this research, the speed of sound velocity in diesel fuel was measured as a function of the temperature between 0 and 35°C. The speed of sound was measured by the clamp on type ultrasonic meter, resulting in the curve fitted Eq. (18) with the coefficients specified by **Table 7**, with a root mean square dispersion of 2.29 m/s (95.45% level of confidence).

$$c_{diesel} = \sum_{i=0}^{2} A_i\, T^i \tag{18}$$

A	B	C	D	E	F
Psi			psi	psi	psi.R^{-1}
1,286,000	13.55	41.220	4530	10.59	3.228

Table 6.
Coefficients for calculating the compressibility modulus by Eq. (17).

A_0	A_1	A_2
1469.2775	−3.1274	−0.01213

Table 7.
Coefficients for calculating the speed of sound in diesel fuel by Eq. (18).

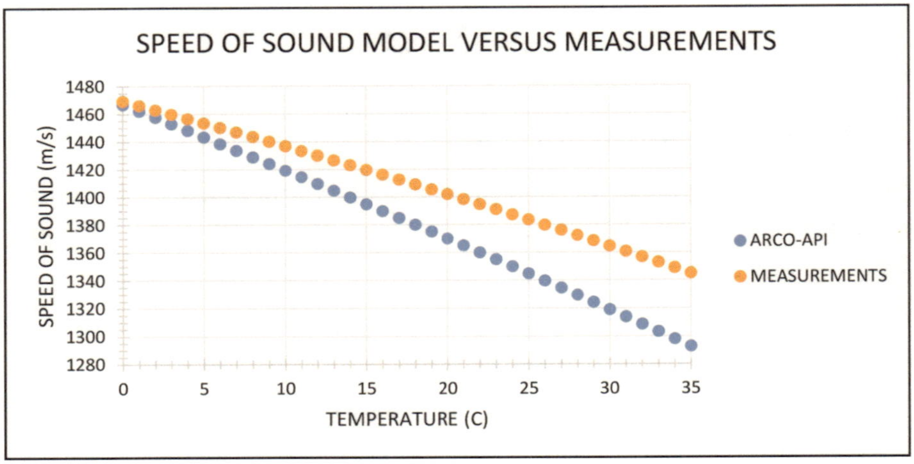

Figure 2.
Comparisons between the speed of sound measured values and model.

The speed of sound measurement values in diesel fuel, Eq. (18) and **Table 7**, was compared with those obtained by the ARCO methodology [13] utilizing the API procedure [12] for calculating the diesel fuel density values as a function of temperature, which is shown in **Figure 2**. The difference is much larger than the stated uncertainty of measurement for water (0.2%).

It is assumed that the uncertainty of measurement of the speed of sound in diesel fuel is the same as in water, considering that the repeatability of the meter for speed of sound measurement is the same for both fluids.

2.6 Temperature

Water and diesel fuel temperatures were measured by two 100 Ω platinum resistance thermometers (Pt100), calibrated with a 25.5 Ω standard platinum resistance thermometer (SPTR) in a constant temperature calibration water bath. The Callendar-Van Dusen equation [14] was utilized to relate the ratio between the measured thermometer resistances at the bath temperature (R) and at 0°C (R_o), and the bath temperature (T), measured by the standard platinum resistance thermometer with an overall uncertainty smaller than 0.05°C. The coefficients of Eq. (19) are shown in **Table 8**.

$$R = R_o\left[1 + A\,T + B\,T^2 + C\,T^3\,(1 - T)\right. \tag{19}$$

During the calibration, the resistances of the standard platinum resistance (SPTR) and the two platinum resistance (Pt100) thermometers were measured at T and 0°C.

A	B	C
0.00390802	−580195E-07	−427350E-12

Table 8.
Coefficients for calculating temperature by Eq. (19) [14].

Pt100 - 1	Pt100 - 2	SPRT
°C	°C	°C
0.00	0.00	0.00
15.17	15.15	15.15
20.18	20.16	20.17
25.18	25.15	25.17
30.17	30.13	30.16
35.05	35.00	35.04

Table 9.
Comparison between the temperature values indicated by the thermometers.

The temperature was calculated, respectively, by the SPTR and by the Callendar-Van Dusen equations. The average measured results are shown in **Table 9**. The difference between the temperature values is smaller than 0.05°C, which gives an uncertainty of 0.1°C (95.45%), when combining the repeatability of the platinum resistance thermometer (Pt100) with the measurement uncertainty of the standard platinum resistance thermometer (SPTR).

3. Experimental methods and data processing

3.1 Experimental facility

Although less accurate, the proposed method can replace the traditional Karl Fisher titration method [15], used in commercially available testing instruments, that requires sampling for laboratory analysis and much time to complete, besides being labor intensive.

The accurate flow rate measurement by the clamp on type meter requires that the user inputs data on the wall material and thickness, besides pipeline diameter. The meter processing software then outputs a value for the speed of sound, based on the measured propagation transit time and calculated acoustic path length, using a proprietary algorithm. The experimental facility, **Figure 3**, consists of a 3″ diameter acrylic tube, with two pairs of piezoelectric sensors mounted on its outer surface to measure the speed of sound in oil–water mixtures, flowing inside by gravity. Both internal and external pipe diameters were measured at five positions along each of the three pipe cross sections, located in a 100-mm-long test section where the sensors were mounted. All measurements were inside the (80.93 ± 0.18) mm range for the internal diameter (87.45 ± 0.18) mm range for the external diameter, and (3.26 ± 0.18) mm range for the wall thickness, which was calculated subtracting the internal diameter from the external diameter.

A previously measured oil–water mixture, starting from pure diesel fuel condition, is stored in RESERVOIR 1, being homogenized by a mixer. After the gravity flow achieves the steady state condition, a clamp on type ultrasonic meter, installed outside a 3″ diameter acrylic tube, measures the speed of sound, registered by a data acquisition system every 6 s. The average value and its standard deviation, which includes both meter measurement characteristics and flow stability, are calculated. The fluid

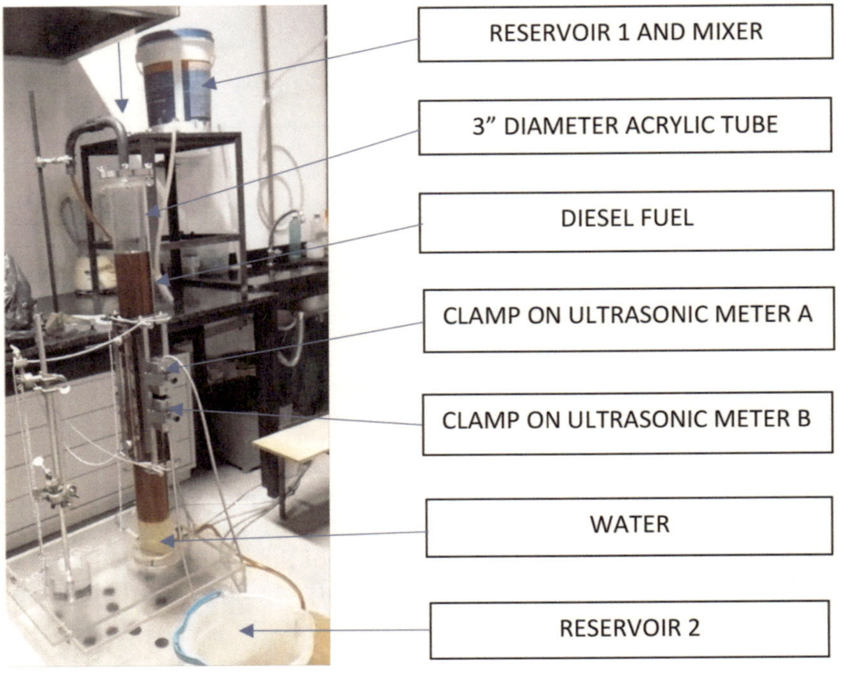

Figure 3.
Experimental facility for measuring speed of sound in oil–water mixtures.

that leaves the test section is stored in RESERVOIR 2, to be used in a next run by adding a measured amount of water to it, thus increasing the water content of the mixture. For large water content values, diesel fuel is gradually added to the mixture, which is initially pure water, thus decreasing its water content.

At the beginning of each run, the amount of oil is weighed and stored in RESERVOIR 1, where the fluid is completely mixed and homogenized for the speed of sound measurement in a steady state gravity flow from the same reservoir. At the end of the run, a weighed amount of water is added to RESERVOIR 2, and the whole amount of fluid is transferred to RESERVOIR 1 for starting a new run. The water content was calculated by Eq. (1), with a smaller than 3 m/s uncertainty of measurement, or 0.20%, which is larger than the uncertainty of curve fitting the measured speed of sound data for diesel fuel (2.29 m/s) and water (1.58 m/s), as a function of temperature.

3.2 Water content estimated from speed of sound measurement

In every run, the speed of sound in the oil–water mixture was measured, together with its temperature, used to calculate the diesel fuel density, Eq. (14), and the water density, Eq. (7). The mixture density then was calculated by Eq. (6). The speed of sound in diesel fuel was calculated by Eq. (18). The speed of sound in water was calculated by Eq. (8).

The water content was estimated from speed of sound measurement by Eq. (5) and compared to the directly measured value in the following water content ranges, (a) 0 to 2.5% (0 to 0.025), (b) 5–20% (0.05 to 0.20), and (c) 80–100% (0.80 to 1.00).

3.3 Water content estimated by the curve fitted measured values

An attempt was made to reduce the uncertainty propagation of oil and water properties for estimating the water content. The measured speed of sound in the oil–water mixture was curve fitted directly as a function of measured temperature and water content values, by Eq. (20).

$$c = \left(A_1 + A_2.T + A_3.T^2\right) + \left(B_1 + B_2.T + B_3.T^2\right)f \tag{20}$$

$$A_i = A_1 + A_2.T_i + A_3.T_i^2 \tag{21}$$

$$B_i = B_1 + B_2.T_i + B_3.T_i^2 \tag{22}$$

$$s = \sqrt{\frac{1}{n-6}\sum\nolimits_{i=1}^{n}\left[c_i - \left(A_i + B_i f_i\right)\right]^2} \tag{23}$$

The coefficients are calculated by minimizing the root mean square deviation of the speed of sound (s), Eq. (23), from the curve fitted Eq. (20),

$$\frac{\partial s}{\partial A_1} = \frac{\partial s}{\partial A_2} = \frac{\partial s}{\partial A_3} = 0 \tag{24}$$

$$\frac{\partial s}{\partial B_1} = \frac{\partial s}{\partial B_2} = \frac{\partial s}{\partial B_3} = 0 \tag{25}$$

Which results in the following system of linear equations to be solved.

$$A_1.X_{11} + A_2.X_{12} + A_3.X_{13} + B_1.X_{14} + B_2.X_{15} + B_3.X_{16} = Y_1 \tag{26}$$

$$A_1.X_{21} + A_2.X_{22} + A_3.X_{23} + B_1.X_{24} + B_2.X_{25} + B_3.X_{26} = Y_2 \tag{27}$$

$$A_1.X_{31} + A_2.X_{32} + A_3.X_{33} + B_1.X_{34} + B_2.X_{35} + B_3.X_{36} = Y_3 \tag{28}$$

$$A_1.X_{41} + A_2.X_{42} + A_3.X_{43} + B_1.X_{44} + B_2.X_{45} + B_3.X_{46} = Y_4 \tag{29}$$

$$A_1.X_{51} + A_2.X_{52} + A_3.X_{53} + B_1.X_{54} + B_2.X_{55} + B_3.X_{56} = Y_5 \tag{30}$$

$$A_1.X_{61} + A_2.X_{62} + A_3.X_{63} + B_1.X_{64} + B_2.X_{65} + B_3.X_{66} = Y_6 \tag{31}$$

Where p stands for matrix line number, and q stands for matrix column number.

$$X_{pq} = \sum_{i=1}^{n} f_i^{w}.T_i^{z} \tag{32}$$

$$Y_p = \sum_{i=1}^{n} c_i.f_i.T_i^{r} \tag{33}$$

And the exponents are specified in **Table 10**.

3.4 Uncertainty of estimating the water content

3.4.1 Measuring mass of diesel fuel and water

In the beginning of each run, a weighed amount of diesel fuel ($M_{diesel} \pm U_{diesel}$) is stored in RESERVOIR 1 for speed of sound measurement. Then, the amounts of water

p	q	w	z	r
1, 2, 3	1, 2, 3	0	p + q - 2	p − 1
1, 2, 3	4, 5, 6	1	p + q - 5	p − 1
4, 5, 6	1, 2, 3	1	p + q - 5	p − 4
4, 5, 6	4, 5, 6	2	p + q - 8	p − 4

Table 10.
Exponents of the variables in Eqs. (32) and (33).

are gradually added to the fluid ($m_{water,i} \pm U_{water,i}$), thus increasing the water content of the oil–water mixture (f_m) in run m, with its uncertainty of measurement (U_{f_m}) given by Eq. (39),

$$M_{water,m} = \sum_{i=2}^{m} m_{water,i} \tag{34}$$

$$M_m = M_{diesel} + M_{water,m} \tag{35}$$

$$f_m = \frac{M_{water,m}}{M_m} \tag{36}$$

$$U_{water,m} = \sqrt{\sum\nolimits_{i=2}^{m} (m_{water,i})^2} \tag{37}$$

$$U_m = \sqrt{(U_{diesel})^2 + (U_{water,m})^2} \tag{38}$$

$$U_{f_m} = f_m \sqrt{\left(\frac{U_{water,m}}{M_{water,m}}\right)^2 + \left(\frac{U_m}{M_m}\right)^2} \tag{39}$$

For large water content values, a weighed amount of water ($M_{water} \pm U_{water}$) is stored, in the beginning of each run, in RESERVOIR 1 for speed of sound measurement. Then, the amounts of diesel fuel are gradually added to the fluid ($m_{diesel,i} \pm U_{diesel,i}$), thus decreasing the water content of the oil–water mixture (F_m) in run m, with its uncertainty of measurement (U_{F_m}) given, likewise, by Eq. (40),

$$U_{f_m} = f_m \sqrt{\left(\frac{U_{diesel,m}}{M_{diesel,m}}\right)^2 + \left(\frac{U_m}{M_m}\right)^2} \tag{40}$$

3.4.2 Curve fitting equation for speed of sound with temperature and water content

Eq. (20) is used to calculate the speed of sound as a function of temperature and water content, which is physically more plausible. However, the objective of this research is to estimate the water content from temperature and speed of sound measurement. The water content in Eq. (20) can be rewritten as a function of (c) and (T).

$$f = \frac{c - A}{B} \tag{41}$$

According to Ref. [16], the uncertainty of estimating the water content (s) by Eq. (41), considering the propagation of the speed of sound uncertainty (u_c) and the temperature uncertainty (u_T), can be expressed by Eq. (42)

$$s = \sqrt{\left(\frac{\partial f}{\partial c} u_c\right)^2 + \left(\frac{\partial f}{\partial T} u_T\right)^2} \tag{42}$$

However, additive corrections [16] must be applied to Eq. (41) to take into account the uncertainty of the water content measurement (u_f), Eq. (39), and the root mean square deviation of the measured data from the curve fitted equation (u_{fit}), Eq. (44), that does not represent precisely the physical phenomenon. Thus, the overall uncertainty of estimating the water content (u) becomes

$$u = \sqrt{u_f^2 + u_{fit}^2 + \left(\frac{\partial f}{\partial c} u_c\right)^2 + \left(\frac{\partial f}{\partial T} u_T\right)^2} \tag{43}$$

$$u_{fit} = \sqrt{\frac{1}{n-1} \sum_{i=1}^{n} \left[f_i - \left(\frac{c_i - A_i}{B_i}\right)\right]^2} \tag{44}$$

where A_i and B_i are calculated by Eqs. (21) and (22), respectively. The sensitivity coefficients are calculated by Eqs. (45) and (46), obtained by differentiating Eq. (41),

$$\frac{\partial f}{\partial T} = -\frac{1}{B}\left(\frac{dA}{dT} + \frac{dB}{dT} f\right) = -\frac{(A_2 + 2\,A_3 T) + (B_2 + 2\,B_3 T) f}{(B_1 + B_2.T + B_3.T^2)} \tag{45}$$

$$\frac{\partial f}{\partial c} = \frac{1}{B} = \frac{1}{(B_1 + B_2.T + B_3.T^2)} \tag{46}$$

Finally, the uncertainty of measurement of the water content (U) from the speed of sound and temperature measurements is given by Eq. (47), with a specified level of confidence for ($n-1$) degrees of freedom,

$$U = t\,u \tag{47}$$

4. Results and discussion

4.1 Water content from measuring mass of diesel fuel and water

The uncertainty of measuring the water content in the (0.80 to 1) range, **Table 11**, is the smallest one, followed by the (0 to 0.025) range, **Table 12**, and finally the (0.05 to 0.20) range, **Table 13**. However, the maximum uncertainty is smaller than 0.0025.

Diesel		Water		Diesel + Water		Water Content	
$M_{diesel,m}$	$U_{diesel,m}$	M_{water}	U_{water}	M_m	U_m	f_m	U_{f_m}
G	G	g	g	g	g		
32771.8	1.49	147108.7	1.82	179880.5	2.35	0.8178	0.0000147
32771.8	1.49	147108.7	1.82	179880.5	2.35	0.8178	0.0000147

Diesel		Water		Diesel + Water		Water Content	
$M_{diesel,m}$	$U_{diesel,m}$	M_{water}	U_{water}	M_m	U_m	f_m	U_{f_m}
32771.8	1.49	147108.7	1.82	179880.5	2.35	0.8178	0.0000147
32771.8	1.49	147108.7	1.82	179880.5	2.35	0.8178	0.0000147
32771.8	1.49	155160.4	1.88	187932.2	2.40	0.8256	0.0000145
32771.8	1.49	163480.2	1.94	196252.0	2.44	0.8330	0.0000143
32771.8	1.49	171847.0	1.99	204618.8	2.49	0.8398	0.0000141
32771.8	1.49	180166.3	2.05	212938.1	2.53	0.8461	0.0000139
32771.8	1.49	180166.3	2.05	212938.1	2.53	0.8461	0.0000139
32771.8	1.49	180166.3	2.05	212938.1	2.53	0.8461	0.0000139
25855.0	1.41	147108.7	1.82	172963.7	2.30	0.8505	0.0000155
21006.6	1.33	147108.7	1.82	168115.3	2.25	0.8750	0.0000160
16423.8	1.24	147108.7	1.82	163532.5	2.20	0.8996	0.0000165
11944.9	1.15	147108.7	1.82	159053.6	2.15	0.9249	0.0000170
7344.9	1.05	147108.7	1.82	154453.6	2.10	0.9524	0.0000175
4410.0	0.94	147108.7	1.82	151518.7	2.05	0.9709	0.0000178
2941.8	0.81	147108.7	1.82	150050.5	1.99	0.9804	0.0000178

Table 11.
Water content from measured mass of diesel fuel and water, (0.80 to 1) range.

Diesel		Water		Diesel + Water		Water Content	
M_{diesel}	U_{diesel}	$M_{water,m}$	$U_{water,m}$	M_m	U_m	f_m	U_{f_m}
G	G	g	g	g	g		
13873.5	1.41	34.8	0.47	13908.3	1.49	0.0025	0.00003
13873.5	1.41	69.7	0.66	13943.2	1.56	0.0050	0.00005
13873.5	1.41	104.8	0.81	13978.3	1.63	0.0075	0.00006
13873.5	1.41	140.0	0.94	14013.5	1.69	0.0100	0.00007
13873.5	1.41	175.5	1.05	14049.0	1.76	0.0125	0.00007
13873.5	1.41	211.2	1.15	14084.7	1.82	0.0150	0.00008
13873.5	1.41	247.2	1.24	14120.7	1.88	0.0175	0.00009
13873.5	1.41	283.1	1.33	14156.6	1.94	0.0200	0.00009
13873.5	1.41	319.3	1.41	14192.8	1.99	0.0225	0.00010
13873.5	1.41	355.7	1.49	14229.2	2.05	0.0250	0.00010
13873.5	1.41	429.0	1.56	14302.5	2.10	0.0300	0.00011
13873.5	1.41	503.2	1.63	14376.7	2.15	0.0350	0.00011
13873.5	1.41	578.0	1.69	14451.5	2.20	0.0400	0.00012
13873.5	1.41	653.7	1.76	14527.2	2.25	0.0450	0.00012
13873.5	1.41	731.1	1.82	14604.6	2.30	0.0501	0.00012

Table 12.
Water content from measured mass of diesel fuel and water, (0 to 0.025) range.

Diesel		Water		Diesel + Water		Water Content	
M_{diesel}	U_{diesel}	$M_{water,m}$	$U_{water,m}$	M_m	U_m	f_m	U_{f_m}
G	G	g	g	g	g		
10082.2	1.49	530.5	1.76	10612.7	2.30	0.0500	0.00017
10082.2	1.49	643.3	1.82	10725.5	2.35	0.0600	0.00017
10082.2	1.49	758.6	1.88	10840.8	2.40	0.0700	0.00017
10082.2	1.49	876.4	1.94	10958.6	2.44	0.0800	0.00018
10082.2	1.49	997.0	1.99	11079.2	2.49	0.0900	0.00018
10082.2	1.49	1120.3	2.05	11202.5	2.53	0.1000	0.00018
10082.2	1.49	1374.2	2.10	11456.4	2.57	0.1200	0.00019
10082.2	1.49	1641.2	2.15	11723.4	2.62	0.1400	0.00019
10082.2	1.49	1919.7	2.20	12001.9	2.66	0.1599	0.00019
10082.2	1.49	2213.4	2.25	12295.6	2.70	0.1800	0.00019
10082.2	1.49	2520.6	2.30	12602.8	2.74	0.2000	0.00019

Table 13.
Water content from measured mass of diesel fuel and water, (0.05 to 0.20) range.

4.2 Procedure for estimating the water content of diesel fuel-water mixtures

For each of the previously prepared diesel fuel-water mixtures, the measured water content is indicated by **Tables 11–13**, together with its uncertainty of measurement.

A data acquisition system was used to register both temperature and speed of sound of the diesel fuel-water mixture every 6 s for a time interval between 10 and 20 min. The average value and the standard deviation of the registered data were calculated.

The standard deviations of the mean temperature and speed of sound values were calculated as the ratio between the standard deviation of the registered data and the square root of the number of measurements. The uncertainty of the average value was calculated by multiplying its standard deviation by the coverage factor.

Two methodologies were used to estimate the water content of the diesel fuel-water mixture, from the mean speed of sound and temperature measured values in the mixture.

- Theory. The water content in each run was calculated by Eq. (5). Water and diesel fuel density values were calculated, respectively, by Eqs. (7) and (14). Water and diesel fuel speed of sound values were calculated, respectively, by Eqs. (8) and (18). The mixture density was calculated by Eq. (6).

- Curve fit. The water content in each run was calculated by Eq. (41), using the mean speed of sound measured value of the mixture, with A and B coefficients obtained from the curve fitted in Eq. (20).

The water content values calculated by the two methodologies were compared with the calculated value by mass measurement and indicated by **Tables 11–13**. The prediction error can be defined as the difference between the estimated value by each methodology and the calculated value by mass measurement, which is, in principle, the more accurate value.

At this point, it is important to observe that the calibration results are valid for estimating the measure and if both calibration and measurement conditions are similar. In principle, if the standard deviation of the water content is determined from calibration, the number of measurements will determine the uncertainty of estimating its average value.

In this calibration, a large number of measurements, registered by the data acquisition system, were made to minimize the influence of the meter repeatability on the results. This procedure can be considered a calibration of a meter for estimating the water content by the speed of sound and temperature measurements, in comparison with a known value of the water content, as determined from mass measurement with very low uncertainty. That is why the systematic error of the speed of sound measurement was not taken into account. Low uncertainties of measurement require individual meter calibration.

4.3 Curve fit methodology

An equation relating water content with speed of sound and temperature can be obtained by curve fitting the measured values of speed of sound and temperature with the water content value calculated from mass measurement. The coefficients of Eq. (20) are shown in **Table 14** for each water content range.

4.4 Comparison between theory and curve fit methodologies

Figures 4–6 show that the utilization of the theory, Eq. (5), for estimating the water content, leads to a large prediction error (< 0.014) in comparison with the value obtained by mass measurement, because of the propagation of the uncertainties of density and speed of sound values in diesel fuel and water. The advantage of using the curve fit methodology, relating directly the water content to the measured temperature and speed of sound of the mixture, is to avoid this propagation, thus reducing the uncertainty. **Figures 4** and **5** show that the prediction error is smaller than 0.002. This procedure can be interpreted as the calibration of the measuring system for water content measurement.

Coefficients	Water Content Range		
	0.000 to 0.025	0.050 to 0.200	0.800 to 1.000
A_1	1604.091911	−3023.601563	−79270.875
A_2	−13.44185019	360.652832	6442.71875
A_3	0.18571924	−7.370956421	−128.722168
B_1	−15240.68262	27805.875	98135.25
B_2	1206.426758	−2238.5	−7834.984375
B_3	−23.64991617	45.15942383	156.6083984

Table 14.
Coefficients of Eq. (20) for each water content range.

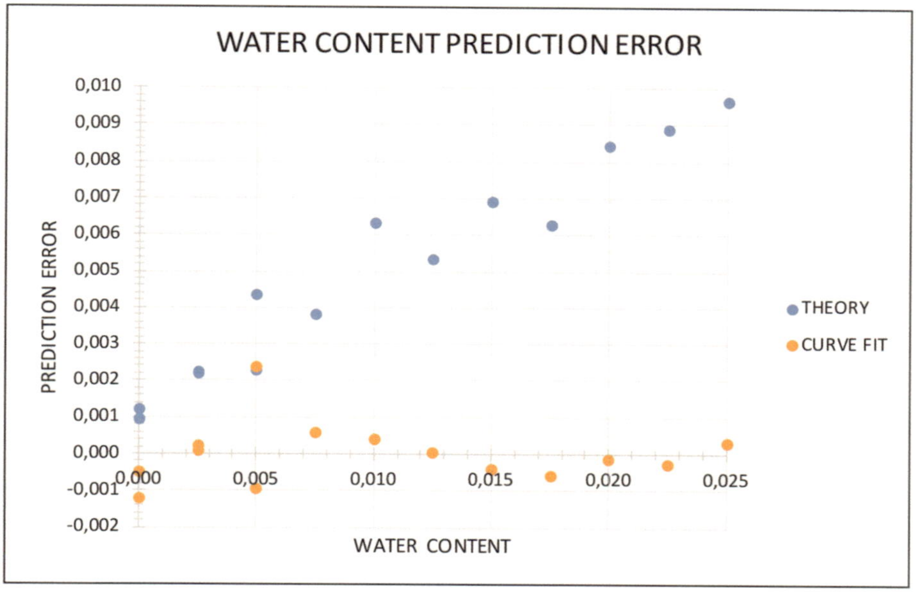

Figure 4.
Water content prediction errors for two methodologies, (0 to 0.025) range.

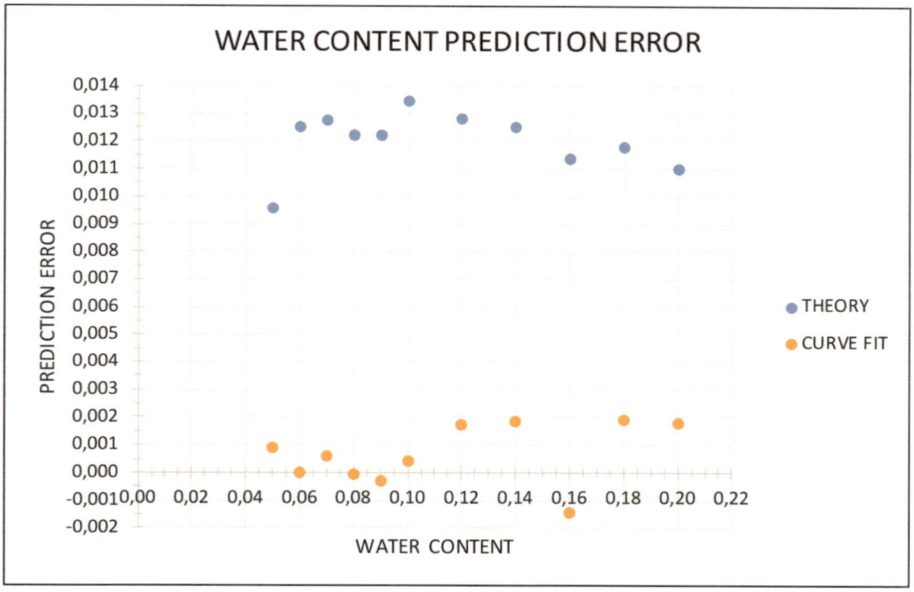

Figure 5.
Water content prediction errors for two methodologies, (0.05 to 0.20) range.

4.5 Uncertainty of measuring the water content by the curve fit methodology

Eq. (43) was used to estimate the uncertainty of water content measurement in different ranges.

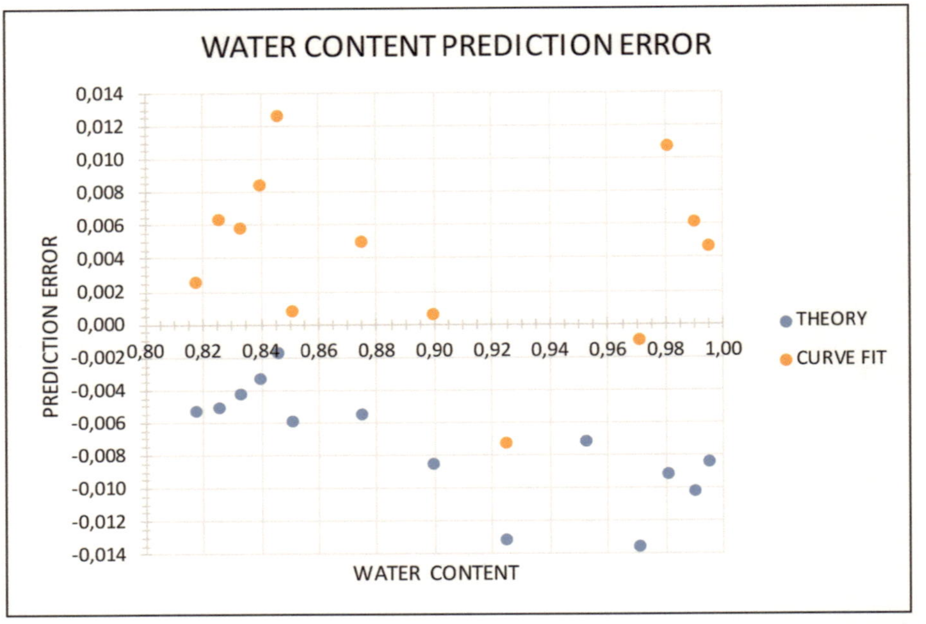

Figure 6.
Water content prediction errors for two methodologies, (0.80 to 1) range.

Table 15 shows that the uncertainty of water content measurement, in the (0 to 0.025) range, is less 0.0025 (0.25%), well inside the accepted limit for fiscal measurement (< 1%).

F	u_f	u_{fit}	u_T	$\partial f/\partial T$	u_c	$\partial f/\partial c$	U
0.0000	0.000017	0.0009	0.00845	0.0271	0.0329	0.0069	0.0020
0.0000	0.000017	0.0009	0.00366	0.0272	0.0130	0.0072	0.0019
0.0025	0.000017	0.0009	0.00365	0.0286	0.0126	0.0076	0.0019
0.0025	0.000024	0.0009	0.00034	0.0292	0.0065	0.0078	0.0019
0.0050	0.000029	0.0009	0.00284	0.0310	0.0148	0.0081	0.0019
0.0050	0.000034	0.0009	0.00328	0.0277	0.0140	0.0071	0.0019
0.0075	0.000037	0.0009	0.00157	0.0297	0.0061	0.0075	0.0019
0.0100	0.000041	0.0009	0.01349	0.0369	0.0544	0.0096	0.0025
0.0125	0.000044	0.0009	0.00595	0.0285	0.0186	0.0070	0.0019
0.0150	0.000047	0.0009	0.01331	0.0294	0.0548	0.0080	0.0023
0.0175	0.000050	0.0009	0.00375	0.0301	0.0114	0.0072	0.0019
0.0200	0.000052	0.0009	0.00086	0.0268	0.0090	0.0075	0.0019
0.0225	0.000055	0.0009	0.00088	0.0267	0.0078	0.0080	0.0019
0.0250	0.000057	0.0009	0.00086	0.0262	0.0066	0.0084	0.0019

Table 15.
Uncertainty of water content measurement, (0 to 0.025) range.

F	u_f	u_{fit}	u_T	$\partial f/\partial T$	u_c	$\partial f/\partial c$	U
0.050	0.000022	0.0013	0.0003	0.1105	0.0398	0.0143	0.0032
0.060	0.000031	0.0013	0.0004	0.0320	0.0549	0.0145	0.0035
0.070	0.000038	0.0013	0.0003	0.0299	0.0610	0.0142	0.0036
0.080	0.000043	0.0013	0.0003	0.0203	0.0564	0.0135	0.0034
0.090	0.000048	0.0013	0.0003	0.0197	0.0573	0.0131	0.0034
0.100	0.000052	0.0013	0.0001	0.0482	0.0403	0.0146	0.0032
0.120	0.000055	0.0013	0.0001	0.0736	0.0629	0.0152	0.0037
0.140	0.000058	0.0013	0.0001	0.0728	0.0646	0.0151	0.0037
0.160	0.000060	0.0013	0.0002	0.0718	0.0843	0.0151	0.0041
0.180	0.000062	0.0013	0.0001	0.0701	0.0854	0.0151	0.0042
0.200	0.000064	0.0013	0.0002	0.0670	0.0531	0.0151	0.0035

Table 16.
Uncertainty of water content measurement, (0.05 to 0.20) range.

Table 16 shows that the uncertainty of water content measurement, in the (0.05 to 0.20) range, is less than 0.005 (0.50%).

Table 17 shows that the uncertainty of water content measurement, in the (0.80 to 1) range, is less than 0.020 (2%). It can be seen that the curve fit in this range is much worse than in other ranges. The probable reason is that the experimental data were taken in a forced flow loop, less stable than the gravity flow loop used for other ranges.

F	u_f	u_{fit}	u_T	$\partial f/\partial T$	u_c	$\partial f/\partial c$	U
0.818	0.000007	0.0081	0.0007	−0.0204	0.137	0.0071	0.018
0.826	0.000007	0.0081	0.0011	−0.0116	0.092	0.0063	0.018
0.833	0.000007	0.0081	0.0004	−0.0201	0.121	0.0071	0.018
0.840	0.000007	0.0081	0.0004	−0.0226	0.098	0.0071	0.018
0.846	0.000007	0.0081	0.0003	−0.0272	0.102	0.0070	0.018
0.851	0.000008	0.0081	0.0006	−0.0273	0.168	0.0070	0.018
0.875	0.000008	0.0081	0.0006	−0.0339	0.215	0.0070	0.018
0.900	0.000008	0.0081	0.0007	−0.0359	0.289	0.0070	0.018
0.925	0.000008	0.0081	0.0005	−0.0329	0.393	0.0071	0.019
0.952	0.000009	0.0081	0.0001	−0.0401	0.477	0.0071	0.019
0.971	0.000009	0.0081	0.0013	−0.0668	0.254	0.0069	0.018
0.980	0.000009	0.0081	0.0010	−0.0928	0.356	0.0068	0.019
0.990	0.000009	0.0081	0.0008	−0.1093	0.198	0.0066	0.018
0.995	0.000009	0.0081	0.0009	−0.1324	0.335	0.0064	0.018

Table 17.
Uncertainty of water content measurement, (0.80 to 1) range.

Tables 15–17 show that the root mean square deviation of the measured data from the curve fitted equation (u_{fit}), Eq. (44), is the largest contribution to the uncertainty of estimating the water content of oil–water mixtures, which can be reduced by utilizing other curve fit equations. Also, a reduction of the uncertainty value by a factor of $\sqrt{n}$ can be obtained by replicating the number n of measurement sets.

5. Critical analysis of the methodology

An 8-acoustic path spool type ultrasonic flow meter measures directly the flow rate with an uncertainty that is compatible to what is required for fiscal measurement. Replacing a section of the pipeline by the dimensionally controlled spool, flow rate and speed of sound values can be measured with good repeatability. The meter piezoelectric sensors are in direct contact with the flowing fluid in the pipeline, and the meter is kept fixed during operation, thus preserving its measuring performance for a long time after the initial calibration.

The clamp on type ultrasonic flow meter is mounted on the outer surface of the pipeline and can be placed anywhere along it, thus being a convenient flow control tool. However, its uncertainty of measurement is worse, mainly due the fact that it is difficult to repeat the same mounting conditions, which suggests that it must remain fixed after an *in situ* calibration, for its preservation. The piezoelectric sensors are not in direct contact with the fluid, and the acoustic wave is propagated through the pipe wall before reaching the fluid. Its propagation velocity depends on the wall material. Furthermore, the propagation velocity through the air gap between the sensors and the pipe wall must be considered to reduce the uncertainty of speed of sound measurement in the flowing fluid inside the pipe.

The error of the speed of sound measurement of the flowing fluid inside the pipeline was estimated by calculating the acoustic wave propagation transit times (T) through the air gap, pipe wall, and fluid. The clamp on type meter was operated in the reflection mode, which means that the sound wave emitted by one sensor is reflected by the pipe wall in the opposite side, reaching its pair mounted on the same side. As a consequence, the path length (L) is doubled, thus reducing the uncertainty of speed of sound measurement. Considering a θ = 45° inclination angle of the acoustic wave, the path length can be calculated by Eq. (48). The transit time (T) is calculated by Eq. (49), from the speed of sound (c)

$$L = 2\,x/\cos(\theta) \tag{48}$$

$$T = L/c \tag{49}$$

Table 18 shows the average values of the speed of sound in pipe wall materials (acrylic and steel), two flowing fluids (water and diesel), and air (gap), used to estimate the transit time of the acoustic wave propagation.

Speed of Sound (m/s)				
WATER	**DIESEL**	**AIR**	**ACRYLIC**	**STEEL**
1480	1530	340	1430	5900

Table 18.
Speed of sound in materials for flow rate measurement.

	Dimensions (mm)	Path Length (mm)
Diameter	80.93	228.90
Wall Thickness	3.26	9.22
Air Gap	0.2	0.57

Table 19.
Dimensions and acoustic path length.

Path Length	Transit Time (μs)			
	Water		Diesel	
	Acrylic	Steel	Acrylic	Steel
Diameter	154.7	154.7	169.6	169.6
Wall	6.4	1.6	6.4	1.6
Air Gap	1.7	1.7	1.7	1.7

Table 20.
Transit time (μs) for different wall materials and flowing fluids.

Table 19 shows the pipe dimensions, estimated air gap thickness, and the acoustic path length through flowing fluids, pipe wall, and air gap, calculated by Eq. (48).

Table 20 shows the transit time, calculated by Eq. (49), for the acoustic wave propagation through the flowing fluid, pipe wall, and air gap. Due to the fact that the average flowing fluid speed of sound is calculated as the ratio between the total path length and the total transit time, and the contribution to its error is a function of the transit time through the pipe wall and air gap, unless it is taken into account, as the meter manufacturer usually does. It can be observed that for steel pipelines, the measurement error for measuring the fluid speed of sound is much smaller than for the acrylic pipe of these experiments. That is why the material properties do not need to be known precisely for correction by the meter software. However, the sensor mounting on the outer surface of the pipe needs to be done with care, to improve the repeatability of the speed of sound measurement, because the software does not make any correction.

The mounting repeatability conditions of the clamp on type meter on the pipe wall are probably responsible for the largest spread of the speed of sound measurements, suggesting that the meter calibration must be verified *in situ* with a fluid with known properties and remained fixed there on.

It has been theoretically shown that the water content of oil–water mixtures can be calculated from temperature and speed of sound measurements. However, the uncertainty propagation of both density and speed of sound for both pure oil and water as a function of temperature results in higher uncertainty values for estimating the water content of the mixture. This problem can be solved by curve fitting directly the water content as a function of the temperature and speed of sound. Another problem to be solved is the traceability of the speed of sound measurement, which is not provided by the ultrasonic meter manufacturers and can be expensive for the user. It was shown in this research that this curve fitting procedure can in fact be considered as a meter calibration, comparing its indicated speed of sound with a traceable water content and temperature, even if the speed of sound value is not known precisely. The problem of this approach is that each meter must be individually calibrated, like fiscal measurement flow meters. The objective of this research is to use the measured speed of sound by the already installed ultrasonic flow meter to continuously estimate and monitor the water content of the mixture. When a spool-type intrusive flow meter is used, the installation effects remain fixed. The meter needs calibration for water content measurement only when the flowing oil is changed in the pipeline. It is suggested that a methodology is developed to use fluids with known properties for *in situ* calibration. For example, the existing facilities for proving fiscal flow meters in the field could be used to sample the oil–water mixture and measure its speed of sound in a range covered by adding water to it. For a clamp on type non-intrusive flow meter, the mounting effects may vary according to its position and mounting conditions. It is suggested that they can minimized by keeping it fixed in a chosen position, with frequent *in situ* verifications, after its calibration in a laboratory, under controlled measurement conditions, which is needed whenever the flowing fluid changes. The advantage of this procedure is that it compensates for meter input inaccuracies of dimensions and material properties. In this research, the speed of sound was measured under different flow conditions, including flow rate values and velocity profiles, showing, as the theory indicates, that they have small influence on the accuracy of its measurement. Therefore, the speed of sound can be measured at both static conditions and at any flow velocity, whichever is more convenient. The fluctuation of the speed of sound measurement by the flow meter is due to meter characteristics and to the fluctuating fluid properties. If the flow meter operates under allowable transient conditions, the spread of measurement can be minimized by considering its average value by a data acquisition system, thus minimizing the uncertainty of water content determination. A larger number of measurements may be necessary when the fluctuation increases.

6. Conclusions

A methodology has been developed to calculate the water content of an oil–water mixture from temperature and speed of sound measurements, using previously measured pure water and pure oil properties as a function of temperature. The propagation of the uncertainties in the determination of the properties is reduced by curve fitting directly the measured values of water–oil mixture speed of sound as a function of temperature and water content. An uncertainty value smaller than 0.0025 (95.45%) was obtained in the range of interest to fiscal measurement (< 0.01). The measurement system is calibrated by relating the meter indicated speed of sound to traceable values of temperature and water content of previously prepared mixtures. The system is able to continuously measure and monitor the water content of an oil–water mixture. For increasing the reliability of the results, it is suggested that a methodology be developed to calibrate *in situ* the meter when the flowing oil changes and to verify the meter calibration frequently.

Acknowledgements

To Petrobras (the Brazilian Oil Company), through R&D Projects, for the financial support that made possible the construction and metrological validation of the laboratory facility for flow measurements at Pontifical Catholic University of Rio de

Janeiro. Also, for having provided measurement data from its operational units, so that they could be analyzed and contribute to the recommended procedure for flow rate measurement. To Evemero Callegario de Mendonça for having planned and taken all the experimental data required for this research. To Flaviomar Soares de Souza for having examined carefully the experimental procedure and organizing the data to be processed.

Author details

Alcir de Faro Orlando
Pontifical Catholic University of Rio de Janeiro, Rio de Janeiro, Brazil

*Address all correspondence to: afo@puc-rio.br

References

[1] Zacharias EM. Method of measuring the amount of water flowing in a crude oil pipeline. US Patent 4.656.869. 1987

[2] Alexander JD, Reed PW. Sonic measurement of flow rate and water content of oil-water streams. US Patent 4.080.837. 1978

[3] Reed PW, Alexander JD. Method and apparatus for sonic velocity type water cut measurement. US Patent 4.236.406. 1980

[4] Orlando AF, Mendonça EC Ferreira MO, Souza FS, Mesquita R. Development of experimental techniques for technical-economical optimization of liquid volume measurement process in tanks and pipelines. Technical Report 5. Petrobras. 2020

[5] Tanaka M, Girard G, Davis R, Peuto A, Bignell N. Recommended table for the density of water between 0°C and 40°C based on recent experimental reports. Metrologia. 2001;**38**:301-309

[6] Greenspan M, Tschiegg CE. Speed of sound in water by a direct method. Research Paper 2795. Journal of Research of the National Bureau of Standards. 1957;**59**(4):249-254

[7] Mackenzie KV. Nine term equation for the sound speed in the oceans. The Journal of the Acoustical Society of America. 1981;**70**(3):807-812

[8] Coppens AB. Simple equation for the speed of sound in Neptunian waters. The Journal of the Acoustical Society of America. 1981;**69**(3):862-863

[9] Chen CT, Millero FJ. Speed of sound in seawater at high pressures. The Journal of the Acoustical Society of America. 1977;**62**(5):1129-1135

[10] Del Grosso VA. New equation for the speed of sound in natural waters (with comparison to other equations). The Journal of the Acoustical Society of America. 1974;**56**(4):1084-1091

[11] Leroy CC, Robinson SP, Goldsmith MJ. A new equation for the accurate calculation of sound speed in all ocean applications. The Journal of the Acoustical Society of America. 2008; **124**(5):2774-2783. Erratum in the Journal of the Acoustical Society of America. 2009;**126**(4):2117

[12] API. Temperature and pressure volume correction for generalized crude oils. refined products and lubricating oils. manual of petroleum measurement standards Chapter 11 – physical properties data. Section 1. 2004

[13] Menon ES. Transmission Pipeline Calculations and Simulations Manual. London, UK: Elsevier Science; 2015 ISBN: 9781856178310 e-Book

[14] IEC 60751. Industrial Platinum Resistance Thermometers and Platinum Temperature Sensors. 2.0 ed. International Electrotechnical Commission International Standard. Geneva, Switzerland. 2008. ISBN: 2831898498

[15] ASTM E203-16. Standard test method for water using volumetric Karl Fischer titration

[16] JCGM 100-2008. Evaluation of measurement data – Guide to the expression of uncertainty in measurement (GUM). Joint Committee on Guides for Metrology (JCGM). 2008

Chapter 5

Perspective Chapter: Geothermics and Thermogenesis in Gas Reservoirs

Yasir Ali and Yasir Yousif

Abstract

Studies on terrestrial heat flow, particularly in oil and gas reservoir systems, have gained substantial attention. While the traditional focus was on igneous and metamorphic activities, this chapter focuses on geothermics and thermogenesis in gas reservoirs, emphasizing the fundamental concepts of heat and temperature, subsurface conditions related to heat, and responses of reservoir materials to temperature changes. Geothermics, at its core, explores the source and destiny of terrestrial heat, with "geo-" denoting the Earth and "thermos" signifying heat. It is the study of heat transport and thermal conditions in the Earth's interior. In practical terms, geothermics extends to the assessment of geothermal resources, examining heat distribution in the Earth's outer layers and the potential for heat extraction. Moreover, this science has evolved into an applied field, with geothermal energy being a notable application that harnesses the Earth's heat. In this context, "thermogenesis" encompasses all physical and chemical reactions in the reservoir, including gas generation, thermal gas cracking, and mineral alteration. In essence, this chapter delves into the intricate dynamics of heat and temperature within gas reservoirs, providing valuable insights into geothermics and thermogenesis, and their significance in the energy industry.

Keywords: earth's heat, heat flow, oil and gas reservoir, geothermic, thermogenesis, thermal gas cracking

1. Introduction

Since its early formation, the earth planet has evolved thermally, and is layered according to density. Earth is subjected to both internal and external source of heat, from the sun and from subsurface. Other source of heat on earth is the heat that originates from the subsurface, which is known later as the ground heat, which is quite important for life on earth, as well as for the earth itself. Volcanoes and seismic activity are induced by such ground heat. Formation of the rocks that form the lithosphere would not have resulted without the interaction between earth matter and earth temperature [1]. One important effect of heat is that the landscape and earth morphology depend to a large extent on internal as well as external heat or thermal stress. In the case of the former, the physical weathering process that acts on reshaping the earth surface is induced by earth temperature as expansion and contraction processes are solely temperature-controlled. Indirectly, temperature effect physical weathering (abrasion process) and deposition (eolian deposits) through wind motivation as wind—the second factor in physical weathering—is influenced by temperature differences from one region to another. Even chemical weathering potential in rocks is enhanced by temperature through the catalyzation of the reaction rate, as well as providing extra surfaces for chemical reactions on rock. An account of the possible temperature-induced geomorphic changes in the land surfaces is given in [2].

2. Heat flow and geothermal gradient

The difference between heat and temperature is almost well-established for everybody. Heat is the vital source of energy, while temperature is the measure of the status of bodies in terms of how cold or hot are they. Therefore, units used for temperature are those of energy (Joule, calorie, ... etc.), and for temperature numerical units such as Celsius, Fahrenheit, and Kelvin degrees are applied. The main source of heat on earth's surface is the Sun, through its radiation which is known as solar radiation. The surface of the earth is subjected to solar radiation during the daytime, and this radiation is lost at night in a continuous reversible process. Another characteristic of such radiation is that it varies over almost all time scales, from daily, through monthly, annually, to century. Spectral composition of this radiation showed that the solar radiation "falls into visible short-wave part of the spectrum, while the other half is mostly in the near-infrared part with a small part" [3]. As geothermal gradient describes the variation in earth's temperature with depth, different formulas have been used to find temperature at any depth based on the gradient/slope. The most famous formula for finding out temperature at any depth is by adding the surface temperature (or sea bottom temperature in case of offshore temperature) to the required depth multiplied by the geothermal gradient at that region. An example of geothermal gradient is shown in **Figure 1**, at a depth of up to 5 km.

The internal heat of the earth has been observed a long time ago and evidenced by the occurrence of volcanic phenomenon [4]. The internal heat is inferred also by the observation of elevated temperature with depth that is associated with subsurface

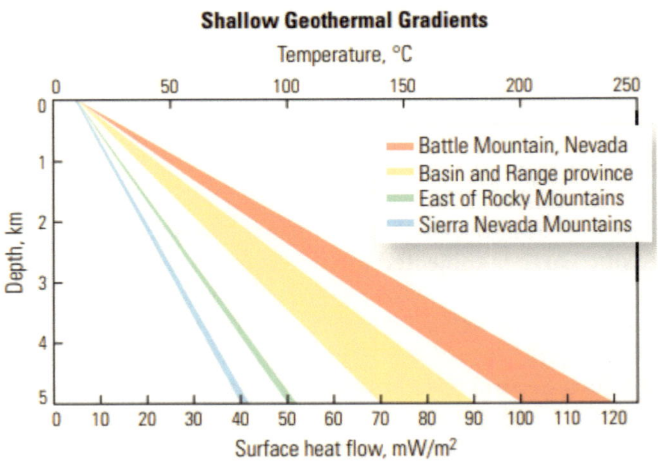

Figure 1.
Shallow geothermal gradient for five different regions. Note variations in surface heat flow based on variations in different geothermal gradients for each region.

drilling for underground mining operations. In such a case, temperature of these tunnels and mines becomes a major problem, which requires the ventilation of the subsurface environment. Trials to understanding this temperature go back to the eighteenth century, when [5] started some discussions on that matter. One important quote of his discussion was reported "because of my being particularly subject to the offended by anything that hinders, full freedom of respiration, I was not solicitous to •oe down into the deep mines"; he, therefore, collected information "(by diligent. inquiry purposely made) from the credible relations of several eye-witnessed suffering in nation, and for the most part unacquainted with each oilier."

However, one cannot talk about the source of earth's heat without talking about the origin of the earth. The most acceptable theory on the formation of the planet state that the earth's matter is condensed by the gravitational force. The earth has become differentiated into different compositional zone-based density, that is, heavy materials sank down to the mantle and lighter materials to the surface forming the earth's crust.

Heat of the earth, which can be classified as external heat and internal is acquired by two means. The exterior heat is acquired during the separation of the earth as young planet, where commits and other floating bodies in the atmosphere hit the earth's surface. The major source of the earth's interior heat is the decay of radioactive minerals [6].

The earth's center is believed to have a temperature of around 6000 C, while the mantle typically ranging from about 1000°C to 3700°C, while the Earth's crust has lower temperatures, ranging from around -40°C to 1000°C, with the highest temperatures in areas with active volcanic activity [7]. Heat or geothermal zonation of the earth is similar to the rock/lithological zonation (**Figure 2**).

The origin of the interior heat is attributed to two main processes that occur inside the earth, namely: radioactivity and earth cooling. Radioactivity contributes 80% of the earth's heat, and heat normally originates as a result of nuclear transformations of radioactive minerals. Radiogenic heat is basically created by the decay of the

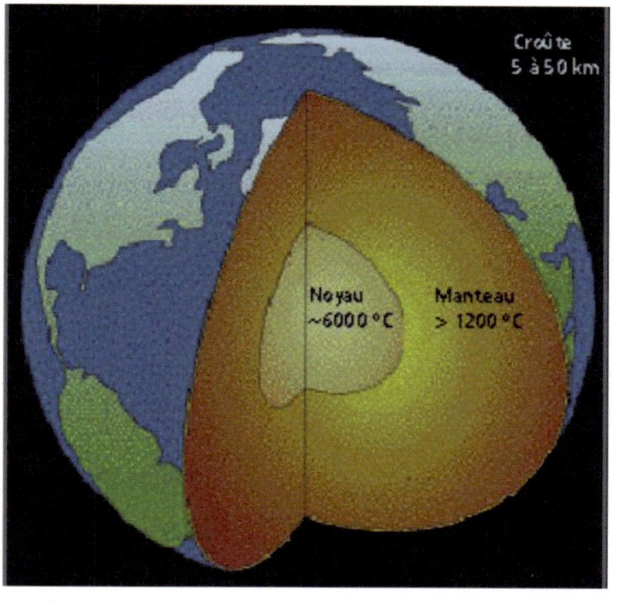

Figure 2.
Geothermal zonation of the earth. Note that the highest temperature is the core (~6000°c) and decreases outward.

three minerals, namely: potassium, thorium, and uranium (K, Th, and U), which accounts for an estimated 30–40% of heat loss through four continents [8–11]. The most famous transformation is that one in which uranium 238 U converts into lead. Radioactive minerals are generally associated with igneous rocks, particularly granitic rocks. Cooling contributes 20% of the earth's interior heat. Radiogenic heat flow is useful in the study of metamorphic rocks and have been applied in the Chinese metamorphic belts. The mean heat production was found to be 0.76 μW m − 3) which is estimated to contribute 24 mW m − 2 to the surface heat flow [12].

Heat production and transport in the earth are illustrated in **Figure 3** for each earth zone. Heat production from radiogenic activity appeared to characterize the

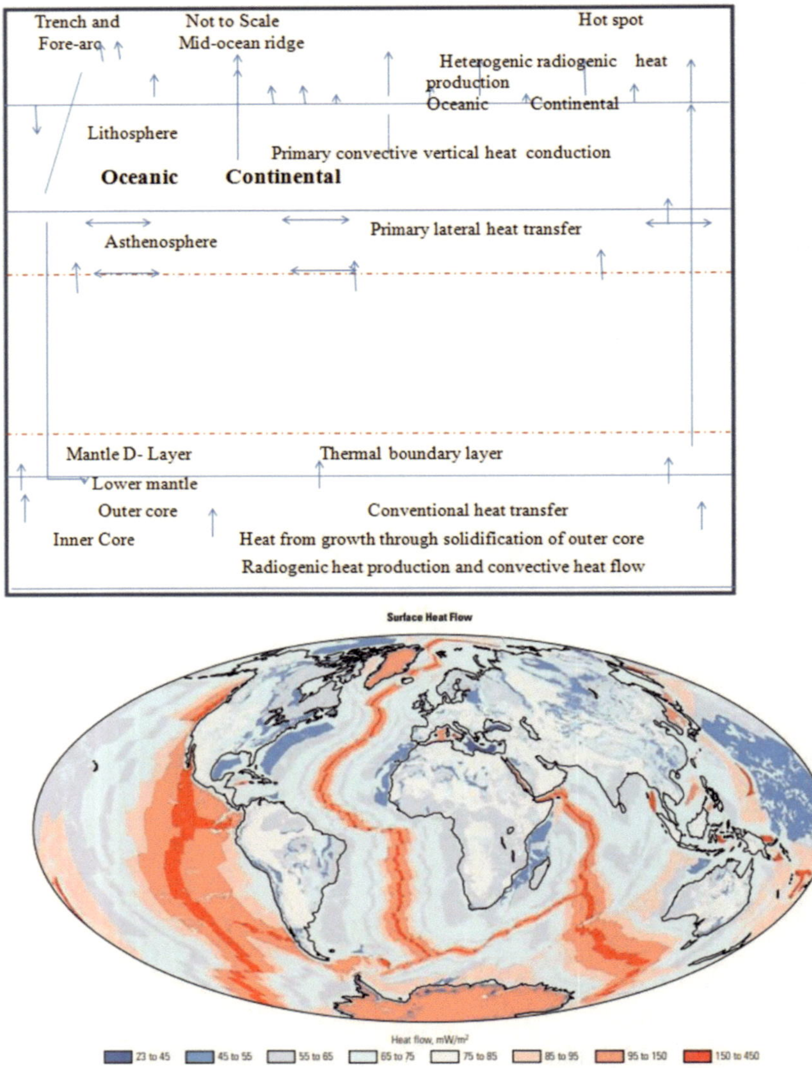

Figure 3.
*Geothermal sources of the earth: A: Heat sources of the earth and heat transport in the earth with respect to different parts of the earth, and b: earth's heat flow from inside to the surface. The color scale (in watts) shows the distribution from the minimal of 23–45 mW/m² (dark blue) to the maximal flux 150–450 mW/m² *reddish).*

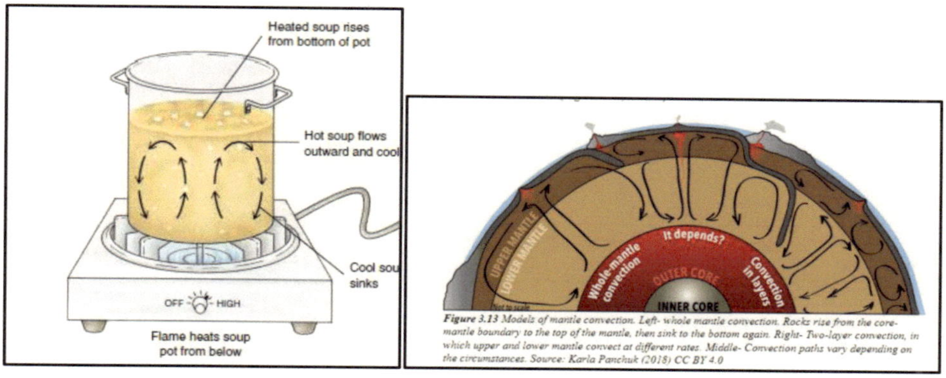

Figure 4.
Demonstration of convection, comparison between soup pot and rock mantle: Soup convects when heated from the bottom of the pot (after [14].

core and lithosphere. Heat due to cooling is referred to as to grow from the solidification of the outer core. Vertical heat transport is shown to be dominant, with minor lateral heat transport. Small portion of the heat generated at the top of the lithosphere is reflected and transferred inward.

As known from fundamental physics, heat is conducted via three modes; namely: radiation, conduction, and convection. In the earth's subsurface, heat moves from the center of the earth outwards through the sedimentary crust into the ocean or atmosphere, where it is lost as radiant energy. The subsurface temperature distribution is influenced by conductive heat transport [13]. Convection, the most important in heat conduction is defined as the process of heat conduction through fluids. The general concept of convection is the movement of material when heated and density lost. The best demonstration for this phenomenon is a comparison an analogy between mantle convection and soup pot, where the content of the soup convects and rises upon heating (**Figure 4**) [14].

3. Thermal properties of rocks

Several recent applications depend on temperature basically induced by thermal properties of rocks; these include high-temperature applications, such as maturation of organic matter for oil generation, oil migration, and enhanced thermal oil recovery, where combustion or hot fluids and steam are injected to the subsurface, geothermal reservoirs, nuclear waste disposal and storage, and groundwater heat storage where hot fluids might be stored and recovered later. On the other hand, there are some processes that are considered low-temperature, such as Perma frost, in the cold regions at low depths, however, these are considered out of the scope of this book. For many constructions, such as underground mines, knowledge of the geothermal gradient is mandatory for the mine design.

There are several thermal properties that are normally considered, including thermal conductivity, heat capacity, and heat diffusivity. Thermal conductivity describes heat flow in steady state flow where no change in temperature with time. In transient heat flow, thermal diffusivity describes the heat flow [15]. Only thermal conductivity and heat capacity will be discussed in the context of thermal imaging heat is conducted from the interior of the earth outward mainly by conduction.

	Thermal conductivity	
Mineral	**W/m-k**	**Btu/ft-hr-F**
Quartz	7.70	4.45
Orthoclase	2.32	1.34
Plagioclase	2.15	1.24
Calcite	3.60	2.08
Muscovite	2.21	1.28
Chlorite	4.91	2.84
Hornblende	3.08	1.78
Epidote	2.61	1.15
Sphene	2.34	1.35
Biotite	2.34	1.35

Table 1.
Thermal of rock-forming minerals after [20].

Heat capacity will be discussed as the process of storing or releasing heat energy will significantly affect the detection of infrared, and hence, the process of resulting thermal imagery.

Thermal conductivity (TC) defines how much heat flows in a rock [16–19]. TC is a vector quantity, unlike density, it depends on the direction of measurement.TC is crucial for the heat flow modeling required for basin thermal history and hydrocarbon generation and migration [13]. Quantification and characterization of heat flow are done through the coefficient TC, which is considered an intrinsic and important petrophysical property.

Thermal conductivity is defined as the capacity of a substance to conduct or transmit heat. This is the coefficient (1) in Fourier's Law of heat conduction:

$$q = -h \, \text{grad} \, T \, (V)$$

where q = heat flux, watts/m2; grad T = temperature gradient, Klm. The standard unit of thermal conductivity is W/m-K. Other units include cakec-cm-"C and Btu/hr-ft-"F.

Thermal conductivity of rocks depend on thermal conductivity of individual minerals constituting that rock. **Table 1** gives values of thermal conductivity of some common rock-forming minerals; in which quartz is the highest conductive mineral and micas are the lowest ones.

4. Heat capacity

Heat capacity is derived from heat content with respect to temperature. Water is taken as the standard material with heat capacity of 1.00 cal/g-"C at 15°C (4.184 kJ/kg-K in SI units), and other substances are compared to that value. Heat capacity is measured experimentally using calorimeters. Review of heat capacity of rocks with respect to their thermal conductivity, it is clear that there is an inverse relationship

Tem (C)	1-Sandstone		2-Sandstone		3-Sandstone		4-Sandstone	
	Exper.	Calc.	Exper.	Calc.	Exper.	Calc.	Exper.	Calc.
127	20.0	19.6	21.3	21.5	21.5	21.9	21.4	21.8
227	42.8	52.5	45.3	45.8	45.7	45.3	45.3	45.1
327	67.3	67.7	71.1	71.4	71.7	72.8	71.3	71.0
427	93.8	93.6	98.6	98.1	99.2	101.0	98.2	98.0
527	120.7	120.0	127.5	126	126.6	130.4	127.7	127.6
	5-Siltstone		6-Siltstone		7-Shale		8-Limestone	
127	21.7	21.3	21.3	21.8	20.9	22.2	22.1	21.5
227	45.8	46.1	46.3	46.0	44/2	44.0	45.8	46.0
327	71.8	71.2	71.7	71.2	69.6	68.4	72.1	71.0
427	99.8	99.4	99.5	97.8	96.3	96.7	98.3	97.3
527	129.0	130.2	127.4	126.5	144.8	124.4	126.2	125.0

Table 2.
Calculated and measured heat content of some sedimentary rocks. Values in cal/g; temperature base 298 k.

between both properties, that is, the higher the thermal conductivity of the rock type, the lower its heat capacity. This is simply interpreted as that conductive rocks cannot keep heat, however, the range of variation in the heat capacity of rocks is not as wide as in thermal conductivity. **Table 2** show heat capacity of common sedimentary rock types, including siliciclastic rocks (sandstone, siltstone, and shale), and carbonate ones. Heat capacity property in reservoirs is important in storing thermal energy by injecting fluids and restoring them later when needed.

Heat capacity for different rock types is proved to be of almost similar values with no signficat variations, where the range was from 20 to 22. In higher temperatures, the range is from 120 to 130 J/g.

5. Thermal diffusivity

Thermal diffusivity describes the heat flux inside a certain volume of the material, while the out flux between the rock and surrounding is the thermal conductivity. In other words, thermal diffusivity controls the rate at which temperature rises inside a uniform block of the material. If, on the other hand, heat capacity reflects the stored heat in a volume that causes the rapid increase of its temperature, we can find that there is a genetic relation between the three rock properties, where rock thermal diffusivity is the ratio of thermal conductivity to heat capacity [21]. The thermal diffusivity is demonstrated in **Figure 5**.

6. Reservoir geothermics

As reservoirs are basically a system of rocks, pore systems, and fluids, the heat flow in such systems is complicated and is the resultant of all components of the system. This section discusses the subsurface conditions in relation to heat and temperature.

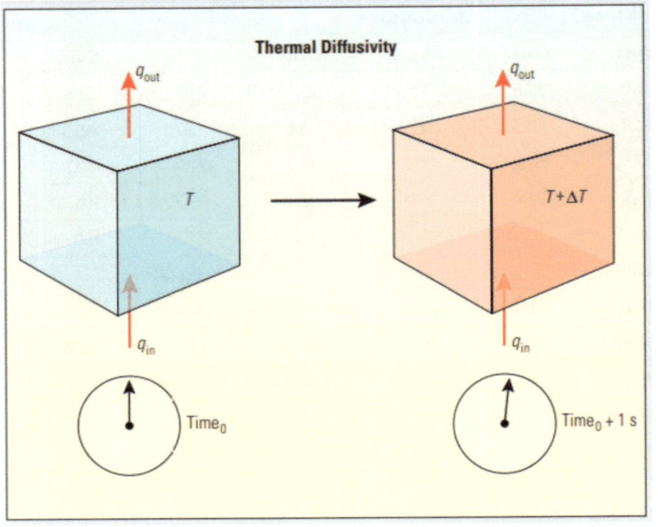

Figure 5.
Demonstration of thermal diffusivity using a block of uniform material.

7. The subsurface PT conditions

In reservoirs, the subsurface is composed of the rock material as a framework, subsurface fluids, and the acting processes such as overburden pressure and temperature. With increasing depth, the pressure, temperature, and salinity increase. Many processes are temperature dependant as well as pressure dependant. Temperature-dependant processes are affected basically by the pre-mentioned thermal properties of reservoir rocks such as thermal conductivity, heat capacity, and thermal diffusivity. Such processes include thermal oil recovery, geothermal reservoirs,

Formation pressure is defined as that pressure other than hydrostatic pressure [22]. Among other factors, including the concentration of salts in formation water, subsurface heat is a major factor in increasing formation pressure. The Daltons law relates temperature to pressure in pressurized systems. In the subsurface, the same law applies where overburden pressure increases with an increase in reservoir temperatures. This relationship is shown in the pressure-temperature-density diagram **Figure 6**.

The figure shows that with increasing depth, temperature and overburden pressure increase.

Thermal conductivity of reservoir rocks is measured for dry, solvent-saturated, and brine saturated to simulate thermal conductivity of the reservoir system. All results showed that TC of brine-saturated sandstone is highest, followed by the solvent-saturated, and the lowest value recorded is for the non-saturated ones. This agrees with the models that consider TC of rocks as a summation of the individual minerals that makeup the rock, and the thermal conductivity of brine is higher than solvent, which is higher than air. Within the saturated sandstone samples, TC recorded for the medium-grained samples show relatively higher values as compared to coarse ones (**Figure** 7).

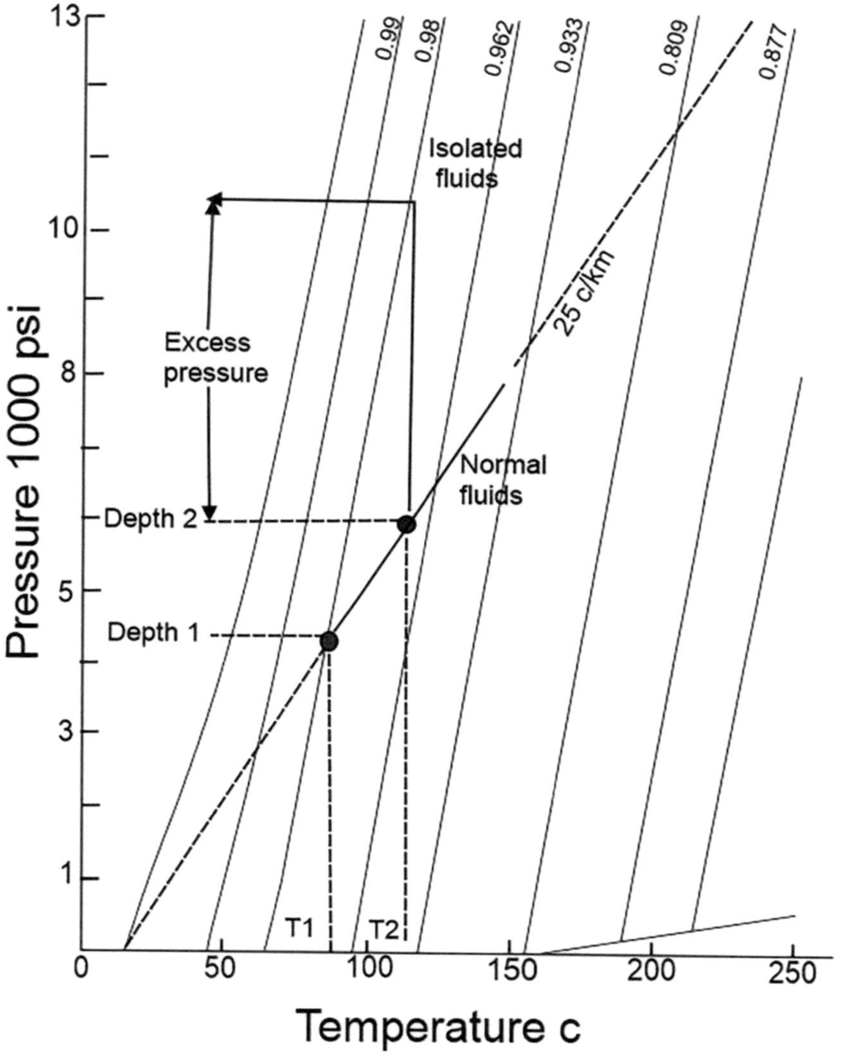

Figure 6.
Pressure-temperature-density diagram for water after [23]. Excess pressure is the higher pressure due to the temperature increase from T 50 to T 60.

In low permeability reservoirs or tight reservoirs, permeability is minimal and fluid may be considered as stationary. However, fluids normally flow under different mechanisms, of which fluids under thermal conditions are to be considered. One of the flow motions is simple convection where fluid moves under temperature/density gradient where hot fluids move upward, while the cold one moves down. Another effect occurs under high temperature where fluid evaporates and the vapor escapes to colder zones where it condenses again. Further details are present in Ref. [18], and initial tests and observations on the effect of vapor pressure and partial saturation of rocks on thermal conductivity were conducted by Ref. [24].

One of the expected subsurface conditions related to temperature is the rock expansion. Study of expansion phenomenon was done for rock forming mineral by

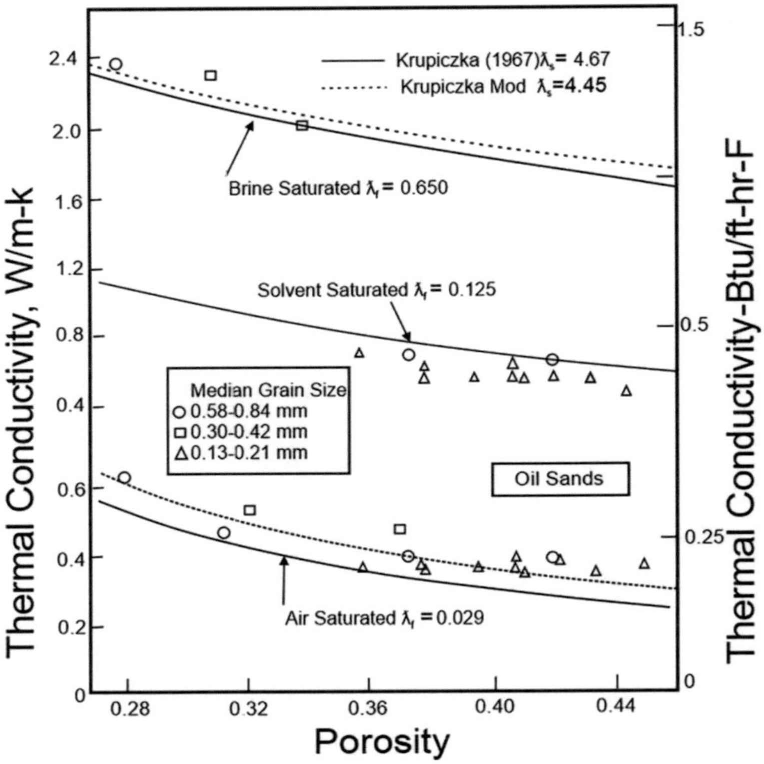

Figure 7.
Thermal conductivity of air-saturated, solvent saturated, and brine-saturated sandstones of oil sands (note the distribution of the points scattered around the curve for each type, where medium-grained ones mostly above the curve, being the highest values for TC.

Ref. [25]. Where gradual expansion with increasing temperature was found to reach up to 2% along the crystal axis. Dry rock expansion was documented by Ref. [26] where direct expansion in rocks was reported as in all natural materials. In saturated rocks, tedious and intensive experiments have been conducted by Ref. [27] who measured experimentally the strain of fluid-saturated rocks using highly sensitive equipments. They concluded that thermal stress on saturated rocks under temperature conditions similar to the subsurface results in the contraction of pore space and increase in fluid expansion. Dry rock expansion, and the comparison of dry sandstone and saturated sandstone is shown in **Figure 8**.

8. Reactions and alteration induced by temperature

Temperature results into thermal stress in almost most materials including earth materials. The response of rocks to thermal expansion is a reflection of responses of the minerals that make up the rock. Mineral alteration by temperature is a well-known

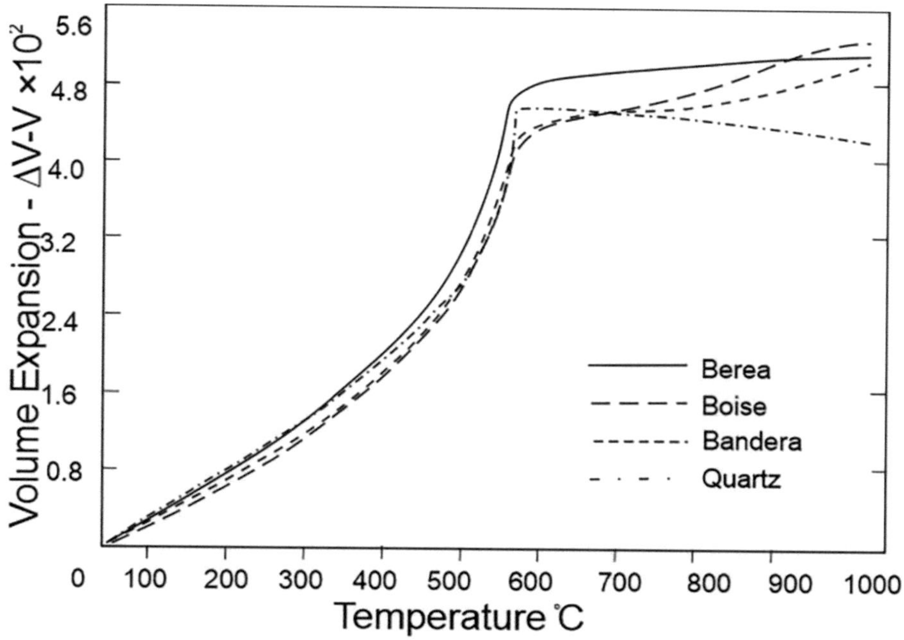

Volumetric thermal expansion of three sandstones compared with quartz

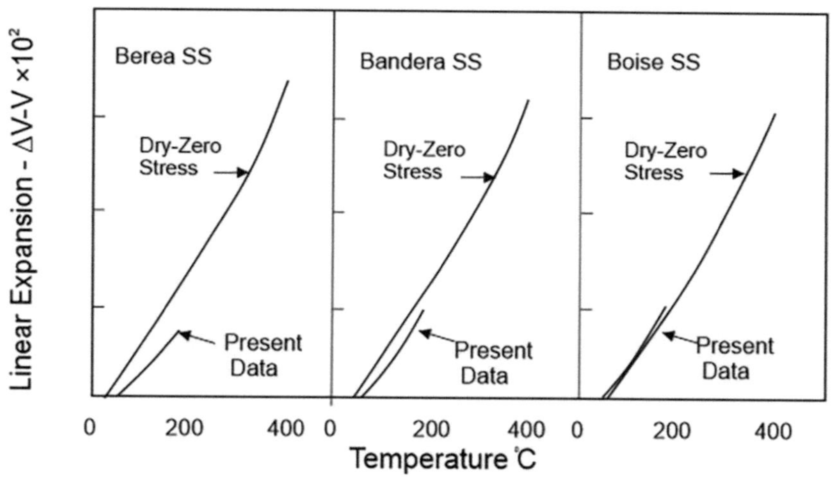

Comparison of linear thermal expansions of dry sandstones and water -saturated sandstones under stress.

Figure 8.
Thermal expansion of rocks: A: Volumetric thermal expansion of three dry sandstone rocks, and b: Thermal expansion of saturated sandstone rocks versus dry ones.

phenomenon as temperature damages the mineral structure through the differential thermal expansion of the minerals. The thermal expansion of minerals varies from one mineral to another where the term coefficient of expansion appears. The coefficient of expansion not only varies from one mineral to another, but it also varies depending on the crystallographic direction.

In most cases, the effects of temperature are evidenced by an alteration of the mechanical properties and rheological behavior of rocks by making fractures similar to that of cooling joints in igneous rocks at the surface. The phenomenon of making fractures in reservoir rocks is in favor of thermal fracing of reservoir rocks in enhanced oil recovery. However, according to [28], the effect of high overburden pressure inhibits the thermal fracking of rocks due to the high weight rocks. Experiments on coalbeds regarding the impact of thermal cracking on the formation of artificial cracks [29] showed that thermal cracks formed proportional to thermal stress in terms of crack size. According to the above-mentioned two cases, the lightweight of coalbeds might make thermal frocking possible as compared to thermal fracking of siliciclastic reservoir rocks.

Diagenesis is another phenomenon that is significantly affected by temperature. In formations under trapped radiogenic heat and high pressures cause diagenesis. Of montmorillonite which decomposes into illite. The former contains compositional water which is released as released freshwater of crystallization either remains in the transformed clay under high pressure because the adjacent sand beds are already geopressured or flows to and dilutes normally pressured aquifers.

The effect of heat in decreasing the viscosity of oil, and subsequently in its movement and migration, is documented in many literature sources; however, scanning electron microscopy techniques were used to demonstrate this effect are shown in **Figure 9**.

While the mantle typically ranging from about 1000°C to 3700°C, while the Earth's crust has lower temperatures, ranging from around –40°C to 1000°C, with the highest temperatures in areas with active volcanic activity [30].

8.1 Occurrence of gas in a gas reservoir

Excluding the secondary occurrence of hydrocarbons in fractured igneous rocks [31], all geologists agree that hydrocarbons do not form in igneous or metamorphic zones, but are generated and retained in sedimentary rocks [32].

The temperature range of formation of both oil and gas from organic matter is called the oil and gas window, respectively. The oil window normally takes place between 60 and 120 ~ gas generation occurs between about 120 and 220 ~ above which the kerogen has been reduced to inert carbon (**Figure 10**).

The subject of oil source rocks is covered in far greater depth in the textbooks [32, 34]. Oil generation normally takes place between 60 and 1R mineral reactions.

Some mineral reactions are generally catalyzed by temperature where they absorb heat. For reversible reactions, absorbed heat is released again and the mineral phase is restored [27, 35]. Experiments have been conducted to demonstrate such reactions by increasing the temperature to simulate the subsurface reservoir (**Table 3**).

9. Reservoir thermogenesis

The fact that source rock evolves thermally is well established [34]. Because of the fact that hydrocarbons in normal cases migrate upward [32], the source rock is subjected to heat flow firstly; and the hydrocarbons migrated from affected shales

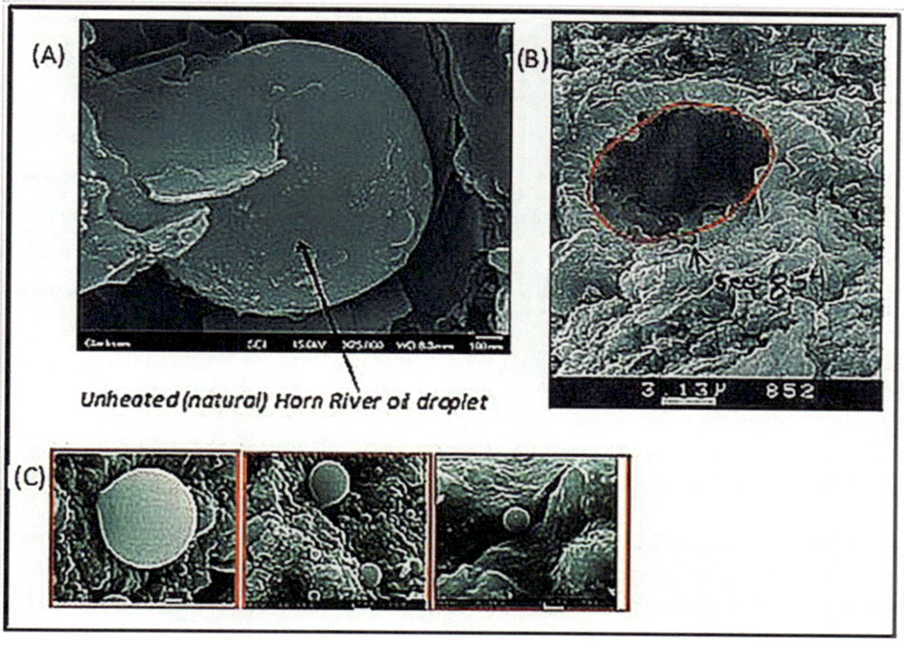

Figure 9.
Conventional SEM images of residual oil droplets in the (a) horn river. (B) Eagle ford, and (C) Woodford shales. The oil migrated into matrix pores, and microfractures upon heating to 350°C for four days in hydrous pyrolysis apparatus (experiments were only run on the Eagle Ford and Woodford formations). The Horn River example, which was not heated, demonstrates that hydrocarbons can occur within matrix mineral pores and not be solely confined to organopores [30].

Figure 10.
Photomicrograph of a heat-affected lower Permian Barakar formation shale from Raniganj basin, India, showing the development of bireflectance in vitrinite [33].

Tem range (C)	Mineral	Heat of reaction cal/g	Reaction
25–220	Ca-Montmorellonite	127	Desorption
25–220	Mg-Montmorellonite	135	Desorption
400–625	Mg-illite	64	Decomposition
455–642	Kaolinite	253	Decomposition
455–723	Ca-Montmorellonite	67	Decomposition
573	Quartz	4.82	α- β inversion
700–830	Ca-carbonate	465	Decomposition
790–950	Mg-illite	15	Decomposition
816–908	Ca-Montmorellonite	26	Decomposition

Table 3.
Heats of reaction for several minerals (after Barshad 1972, [36]).

to reservoir. The abnormal heat flow is generally a heat plume resulting from intrusion. In such cases, hydrocarbon matter volatiles to the reservoir and vitrinite in the affected shale reflects birefringence as shown in the photomicrograph by [33]. However, maturity of hydrocarbons in a reservoir might not reflect the thermal status of the source rock, where immature oils can occur in reservoirs of thermal history, and vice versa. For gas formation, gas is formed either by the effect of bacteria, the process known as biogenesis, at relatively shallower depth below 550 C [37]; or, at higher depths, is formed by the effect of temperature, known as thermogenic gas. In unconventional petroleum systems where the source rock is the reservoir itself, the thermal effect causes the formation of secondary porosity [38]. Such secondary porosity significantly increases the storage capacity of the system.

Geothermal reservoirs represent one of the applications of the responses of reservoir fluids to reservoirs geothermic. Geothermal reservoirs provide clean renewable energy, [39] and the research in that field are going on for decades [40] according to [41] the geothermal system can be defined as a reservoir in a certain area that provides the opportunity to extract heat economically. Generally, the geothermal system can be divided into three main groups [42, 43]. The first one is the hydrothermal system which is formed when heat is transferred from a source by conductivity to porous media and the porous fluid within it. Moreover, the hydrothermal system can be classified into liquid-dominated and vapor-dominated (or dry steam/steam-alone) systems, depending on the existence of water or vapor. The second one is the Geopressured-geothermal system formed when the water is trapped in permeable media (rock) surrounded by impermeable or low permeable rock [42]. The last one is the artificial geothermal reservoir system called the hot dry rock system (HDR), formed in which boreholes are drilled and water is injected into the hot igneous rock [41]. Furthermore, the geothermal systems can be classified upon the equilibrium state into static and dynamic systems, the static is characterized by contentious recharge and discharge of water, while the dynamic is dominated by low or no recharge [42].

In terms of temperature conditions, or reservoir geothermic, geothermal reservoirs are classified into low, medium, and high geothermal reservoirs [41, 42, 44, 45]. The low-temperature geothermal reservoirs (at temperature less than 90°C) can provide waters source for industrial use and other uses but are not sufficient for electricity [44]. The intermediate temperature (the temperature is <150°C and ≥ 90°C), and the high temperature (the temperature is ≥150°C). More details on the subject matter can be found in [46].

10. Thermal fracking

One of the thermal applications of reservoir geothermic also is thermal oil recovery [47–50]. In conventional hydrocarbon systems, heat is used in oil recovery to enhance the oil recovery by reducing the oil viscosity allowing it easily to flow (Reference). The process depends totally on reservoir thermal properties. However, in unconventional shale gas and shale oil recovery, heat is used as a fracking agent instead of hydraulic fracking which requires injecting a large volume of water [51].

11. Conclusion

The geothermics of reservoir have been studied and investigated where sources of terrestrial heat, means of heat transport, and thermal properties of rocks were discussed in detail and in relation to subsurface conditions such as overburden pressure and fluid content. The subsurface heat and subsequent temperature were proven to affect the reservoir system and many processes that take place in it. Porosity changes by the effect of temperature and similarly, fluid phase changes also. Some minerals show alterations and reactions have been described also. The applications of reservoir geothermics have also been addressed such as geothermal reservoirs. In conclusion, this chapter provides deep insight into the heat regime and temperature in reservoirs and, especially, reactions between different subsurface heat and components of reservoir systems.

Acknowledgements

The authors are willing to acknowledge the Red Sea University as their parent university where several facilities are provided for authoring this chapter.

Conflict of interest

The authors declare no conflict of interest.

Author details

Yasir Ali[1*] and Yasir Yousif[2]

1 Red Sea University, Port Sudan, Sudan

2 Pan African University, Ibadan, Nigeria

*Address all correspondence to: yasirovic2009@gmail.com

References

[1] Bagdassarov N, Bagdassarov N. Fundamentals of rock physics. Cambridge University Press; 9 Dec 2021

[2] Starkel L. The palaeogeography of mid-and East Europe during the last cold stage, with west European comparisons. Philosophical Transactions of the Royal Society of London. B, Biological Sciences. 1977;**280**(972):351-372

[3] Cermak V, Safanda J, Bodri L. Precise temperature monitoring in boreholes: Evidence for oscillatory convection? Part 1: Experiments and field data. Int. Journal of Earth Science. 2008;**97**(2): 365-373. [Internet]. DOI: 10.1007/s00531-007-0237-4

[4] Haenel R, Stegena L, Rybach L, editors. Handbook of Terrestrial Heat-Flow Density Determination: With Guidelines and Recommendations of the International Heat Flow Commission. Springer Science & Business Media; 6 Dec 2012

[5] Wright WF. Early evolution of the thermometer and application to clinical medicine. Journal of Thermal Biology. 1 Feb 2016;**56**:18-30

[6] Stacey FD, Davis PM. Physics of the Earth. Cambridge University Press; 28 Aug 2008

[7] Pahud D. Geothermal energy and heat storage. SUPSI – DCT – LEEE. Laboratorio di Energia, Ecologia ad Economia. 2002;**133**:1-133

[8] Artemieva IM, Mooney WD. Thermal thickness and evolution of Precambrian lithosphere: A global study. Journal of Geophysical Research - Solid Earth. 2001;**106**(B8):16387-16414

[9] Hasterok D, Chapman D. Continental thermal isostasy I: Methods and sensitivity. Journal of Geophysical Research. 2007;**112**:B06414. DOI: 10.1029/2006JB004663

[10] Vitorello I, Pollack HN. On the variation of continental heat flow with age and the thermal evolution of continents. Journal of Geophysical Research. 1980;**85**(B2):983-996

[11] Pollack H, Chapman D. Mantle heat flow. Earth Planet. 1977;**34**:174-184

[12] Lin C-H. Thermal modeling of continental subduction and exhumation constrained by heat flow and seismicity in Taiwan. Tectonophysics. 2000;**324**(3):189-201 [Internet] Available from: https://www.sciencedirect.com/science/article/pii/S0040195100001177

[13] Nadeau PH, Bjørkum PA, Walderhaug O. Petroleum system analysis: Impact of shale diagenesis on reservoir fluid pressure, hydrocarbon migration, and biodegradation risks. In: Geological Society. Petroleum Geology Conference series. Vol. 6. No. 1. London: The Geological Society of London. 2005. pp. 1267-1274

[14] Arevalo R, McDonough WF, Luong M. 3.3 Earth's interior heat. Earth and Planetary Science Letters. 2009;**278**(3-4):361-369

[15] Barkaoui AE, Zarhloule Y, Verdoya M, Pasquale V, Lahrach H. Progress in understanding the geothermal sedimentary basins in northeastern Morocco. Journal of the African Earth Sciences. 2014;**97**:1-8 [Internet] Available from: https://www.sciencedirect.com/science/article/pii/S1464343X14001435

[16] Jessop AM. Developments in solid earth geophysics. Thermal Geophysics. Vol. 17. 1990. pp. 293-295

[17] Clauser C, Huenges E. Thermal conductivity of rocks and minerals. Rock Physics and Phase Relations: A Handbook of Physical Constants. Vol. 1. 1992. pp. 271-302

[18] Balkan E, Erkan K, Şalk M. Thermal conductivity of major rock types in western and central Anatolia regions, Turkey. Journal of Geophysics and Engineering. 1 Aug 2017;**14**(4):909-919

[19] Chang-Ya ZB, Hai S, Yi OB, Yang S, Hai HX, Lin C, et al. Research on tectono-thermal evolutionmodeling methods for superimposed basin - with the Jiangan Basin as an example. Chinese Journal of Geophysics. 2010;**53**(1):92-102

[20] Horai K. Thermal conductivity of rock-forming minerals. Journal of Geophysical Research. 1971;**76**(5):1278-1308

[21] Chekhonin E, Parshin A, Pissarenko D, Popov Y, Romushkevich R, Safonov S, et al. When rocks get hot: Thermal properties of reservoir rocks. Oilfield Review. 2012;**24**(3):20-37

[22] Fertl WH. Abnormal Formation Pressures : Implications to Exploration, Drilling, and Production of Oil and Gas Resources / Walter H. Fertl, with a Contribution by George V. Chilingarian and Herman H. Rieke. Vol. III. Developments in Petroleum Science. New York, NY: Elsevier; 1976

[23] Barker C. Aquathermal pressuring—Role of temperature in development of abnormal-pressure zones: GEOLOGICAL NOTES. American Association of Petroleum Geologists Bulletin. 1972;**56**(10):2068-2071. [Internet]. DOI: 10.1306/819A41B0-16C5-11D7-8645000102C1865D

[24] Anand J. Thermal Conductivity of Fluid Saturated Rocks at Elevated Pressures and Temperatures. MS Thesis. Berkeley: University of California; 1971. p. 104

[25] Clark S Jr. Handbook of physical constants. Revised Ed GSA Memoent. 1966;**97**:459-482

[26] Mitoff S, Pask J. A recording differential expansion apparatus. American Ceramic Society Bulletin. 1956;**35**:10

[27] Somerton WH, Selim MA. Additional thermal data for porous rocks--thermal expansion and heat of reaction. Society of Petroleum Engineers Journal. 1961;**1**(04):249-253

[28] Hill RJ, Tang Y, Kaplan IR, Jenden PD. The influence of pressure on the thermal cracking of oil. Energy & Fuels. 1996;**10**(4):873-882

[29] Ibrahim YA, Padmanabhan E. Thermally - Induced Fractures in Coal. International Journal of Coal Geology. 2019:11. Balingian Province, Sarawak, Malaysia: 2018 AAPG International Conference and Exhibition, Industrial Research

[30] Slatt R, O'Brien N. Variations in shale pore types and their measurement. In: Unconventional Resources Technology Conference, Denver, Colorado. Society of Exploration Geophysicists, American Association of Petroleum Geologists, Society of Petroleum Engineers; 2014. pp. 1181-1187

[31] Lemonnier P, Bourbiaux B. Simulation of naturally fractured reservoirs. State of the art-part 1–physical mechanisms and simulator formulation. Oil & Gas Science and Technology–Revue de l'Institut Français du Pétrole. 1 Mar 2010;**65**(2):239-262

[32] Waples DW. Geochemistry in petroleum exploration. Springer Science & Business Media; 1 Dec 2013

[33] Hazra B, Varma AK, Bandopadhyay AK, Mendhe VA, Singh BD, Saxena VK, et al. Petrographic insights of organic matter conversion of Raniganj basin shales, India. International Journal of Coal Geology. 2015;**150**:193-209

[34] Tissot BP, Welte DH. othersPetroleum Formation and Occurrence. A New Approach to Oil and Gas Exploration. Germany: Book in German; 1978

[35] Kelley KK, King EG. Contributions to the data on theoretical metallurgy XIV. In: Entropy of the Elements and Inorganic Compounds. Vol. 592. U.S. Bureau of Mines Bulletin. U.S. Department of the Interior. Washington: United States Government Printing Office; 1961. pp. 1-149

[36] Michel R, Barshad I. Transformation expérimentale des micas en vermiculites ou smectites. Propriétés des smectites de transformation. Bull du Groupe français des Argiles. 1972;**24**(2):137-151

[37] Shurr GW, Ridgley JL. Unconventional shallow biogenic gas systems. American Association of Petroleum Geologists Bulletin. 2002;**86**(11):1939-1969

[38] Loucks RG, Reed RM, Ruppel SC, Hammes U. Spectrum of pore types and networks in mudrocks and a descriptive classification for matrix-related mudrock pores. American Association of Petroleum Geologists Bulletin. 2012;**96**(6):1071-1098

[39] Washington D. Geothermal Energy | National Geographic Society [Internet]. 2015. pp. 1-4. [cited 2022 Sep 15]. Available from: https://education.nationalgeographic.org/resource/geothermal-energy

[40] Manzella A. Geothermal energy. EPJ Web of Conferencs. 2017;**148**:1-26

[41] Gupta HK, Roy S. Geothermal Energy: An Alternative Resource for the 21st Century. Elsevier; 2006

[42] Ganguly S, Kumar MSM. Geothermal reservoirs - a brief review. Journal of the Geological Society of India. 2012;**79**(6):589-602

[43] Finger JT, Blankenship DA. Handbook of best practices for geothermal drilling. United States; 2012. [Internet]. DOI: 10.2172/1325261

[44] Akar S, Turchi C. Low temperature geothermal resource assessment for membrane distillation desalination in the United States. Transactions - Geothermal Resource Council. 2016;**40**(October):129-140

[45] Wang G, Li K, Wen D, Lin W, Lin L, Liu Z, et al. Assessment of geothermal resources in China. In: Proceedings, Thirty–Eighth Workshop on Geothermal Reservoir Engineering, Stanford University, Stanford, California, February 11. 2013. pp. 11-13

[46] Brown DW, Duchane DV, Heiken G, Hriscu VT. Mining the earth's Heat: Hot Dry Rock Geothermal Energy. Springer Science & Business Media; 2012

[47] Gottfried BS. A mathematical model of thermal oil recovery in linear systems. Society of Petroleum Engineers Journal. 1965;**5**(03):196-210

[48] Hascakir B. Introduction to thermal Enhanced Oil Recovery (EOR) special issue. Journal of Petroleum Science and Engineering. 1 Jun 2017;**154**:438-441

[49] Thomas S. Enhanced oil recovery-an overview. Oil & Gas Science and Technology-Revue de l'IFP. 2008;**63**(1):9-19

[50] Nian Y-L, Cheng W-L. Insights into heat transport for thermal oil recovery. Journal of Petroleum Science and Engineering. 2017;**151**:507-521

[51] Hu L, Li H, Babadagli T, Xie X, Deng H. Thermal stimulation of shale formations by electromagnetic heating: A clean technique for enhancing oil and gas recovery. Journal of Cleaner Production. 2020;**277**:123197

Chapter 6

Coking

Jafar Ramezanzadeh and Hossein Moradi

Abstract

Currently, conventional oil is used as the main source for the petrochemical industry. However, conventional oil's capacity is declining, and that source will probably be exhausted in the near future. Heavy oil and petroleum residues have become a suitable alternative source to meet global energy demand. However, heavy oil and oil residues require many upgrading processes before turning to be valuable products. Among the various upgrading processes, delayed coking, which is capable of processing any residue at a low investment cost, garnered tremendous importance. Petroleum coke is one of the coking products that is divided into three types: shot coke, sponge coke and needle coke, depending on the feed properties and operating conditions of the process. Needle coke is used as a valuable product in the production of graphite electrodes used in electric arc furnace (EAF) for melting scrap metal and producing steel.

Keywords: heavy oil, petroleum residue, upgrading process, delayed coking, needle coke, graphite electrodes, electric arc furnace

1. Introduction

The petroleum industry provides most of the world's energy needs and has been the world's most important energy source since the mid-1950s because of its high energy density, easy transportability and relative abundance [1]. Due to rapid population growth, the consumption of fuels, energy, and petrochemical products has increased sharply [2]. At present, light crude oil reserves are the main source of energy that meets global energy demand due to high quality and low production costs. Nevertheless, light crude oil reserves are declining. Such a rapid decline in light crude oil reserves poses great challenges to meeting the world's energy needs. Over the past few decades, renewable, nuclear and bioenergy have been developing rapidly; however, these resources are costly and insufficient in meeting energy demands, especially for transportation [3]. Therefore, refineries have to depend increasingly on unconventional feedstocks such as heavy oils, oil residues, and bitumen to supply the increasing demand for fuels [1]. The fundamental characteristics of heavy crude oil are low American Petroleum Institute gravity (API), low economic value, high viscosity, and high asphaltenes content which makes it more difficult to transport and process than conventional crude oil [4]. This fact leads to an emphasis on the upgrading of heavy and residual oil. The purpose of upgrading heavy oil and residues is to convert feedstock with high boiling point and low H/C ratio to low boiling point distillate fractions and higher H/C ratio and to eliminate hetero atoms such as sulphur, nitrogen, and metals to Environmentally acceptable levels. To achieve this goal, hydrocarbon molecules are exposed to thermal and catalytic cracking reactions during the upgrading processes [5]. According to the approaches to achieve higher H/C ratios, upgrading technologies can be divided into carbon rejection and hydrogen addition processes. Carbon rejection rejects the carbon into carbonaceous product (coke) to obtain lighter products (with a high H/C ratio) in these processes. On the other hand, hydrogen addition processes such as hydrocracking involve the reaction of raw materials with an external source of hydrogen in the presence of a catalyst, which leads to an overall increase in the H/C ratio [6]. Hydrogen addition processes have higher quality and yield of desired products. However, these processes require the participation of hydrogen and catalysts, which leads to higher investment and operating costs compared to carbon rejection processes. In contrast, carbon rejection processes are superior to hydrogen addition processes in terms of simplicity and operating costs, and therefore have many units in the world [7, 8]. Petroleum residues processing capacity indicates that the major portion (approximately 63 wt.%) of petroleum residues are upgraded by thermal processes such as visbreaking and delayed coking [2].

2. Carbon rejection processes

Carbon rejection technologies have been used by refineries since 1913 to upgrade various hydrocarbon feeds. These technologies include visbreaking, gasification, and coking processes. visbreaking and coking technologies can be generally applied to all residual feeds because they are not limited to constraints such as metal content and coke-forming tendencies as in the case of catalytic processes for upgrading. In a carbon rejection process, the feeds (larger molecules) are heated under inert atmospheric pressure to fracture them into smaller molecules [2, 9, 10].

2.1 Visbreaking

Visbreaking remains the oldest and least costly of the upgrading option and is only used in areas where heavy fuel oil is used to generate electricity and fuel ships. Visbreaking is a process in which residues are slowly cracked to reduce viscosity, and its main product is fuel oil, which has a dwindling market and provides low margins. This is a very low conversion process, and 15–20 wt. % residues are converted into lighter fractions. The yield of gas and gasoline together is generally limited to a maximum of about 7 wt. % as the cracking reactions are arrested to prevent asphaltene flocculation. Current interest in visbreaking is in those areas where motor fuel demand is relatively low. Vacuum residue and atmospheric residue can be used as feedstock for the visbreaking process [2, 6, 8, 10].

2.2 Gasification

The Texaco Gasification Process (TGP) was developed in the late 1940s. This process involves the complete cracking of residues into gaseous products, which has received less attention than other processes. Residual gasification is done at high temperatures (>1000°C) and synthetic gas (hydrogen and carbon monoxide), carbon black, and ash are the major products. It was modified in the 1950s for heavy oil feeds, in the 1970s for solid feeds such as coal and in the 1980s for petroleum coke. Almost from the beginning, this process has been attractive for hydrogen production. gasification can be used by refineries to produce hydrogen, increase the yield of high-value products, eliminate the production of high sulphur fuel oil, minimize the environmental effects of refineries (reduce the emission of NO_x and SO_x pollutants) and process a wider range of crude oil [2, 9, 11].

2.3 Coking

Coking is a process in which raw materials are thermally decomposed into products with lower boiling points. Different types of coking processes include delayed coking, fluid coking, and flexicoking. Delayed coking is the most common technology used in petroleum refineries to produce petroleum coke. More than 90% of petroleum coke is produced by this process. The main reasons are the relatively low investment cost and the claims of a better quality of liquid products compared with the fluid or flexicoking process [12].

2.3.1 Delayed coking

In the delayed coking process, the general goal of such a technology is to maximise liquid product yield while minimising coke production. The inherent flexibility of the delayed coking process for handling various feedstocks gives the refinery a promising solution to the problem of decreasing residual fuel demand and takes advantage of the attractive economics of upgrading it to more valuable lighter products. A refinery with a delayed coker is called a 'zero resid refinery' that can convert various feedstocks to valuable engine fuels while eliminating unsold refinery flows that are environmentally unfriendly. Disadvantages of this technology can be the abundant production of coke, low yield of liquid products, and highly aromatic products which require post-treatment. Another disadvantage of delayed coking is that it is a more expensive process than solvent deasphalting. Environmental pollution from coke particles is also a concern. In this process, 20–30 wt.% coke is also produced as a by-product. Although coke is accepted as a by-product of coking processes, excessive coke formation is economically disadvantageous because the value of coke is much lower than that of distillates. Even considering these disadvantages, delayed coking is the most frequently preferred process for refiners to residue processing because of the low investment cost [2, 6, 8–10].

Delayed coking is a severe form of thermal cracking process that operates at low pressures, without the use of hydrogen and catalysts, and falls in the temperature range of 450–500°C. Delayed coking is highly efficient in rejecting mineral solids and metals as well as some organic nitrogen and sulphur in the coke. The name 'delayed' derives from the fact that cracking reactions are given enough time (long) to form coke in coke drums. The first commercial delayed coker was started in 1930 at Standard Oil's Whiting refinery [12, 13].

The global trend of processing heavy raw materials in delayed cokers, in order to obtain maximum yield of liquid products, has led to the production of coke with fuel grade that contains large amounts of sulphur and metals. Fuel grade coke, once considered a by-product of waste, is now an important fuel for the cement industry and electricity generation [6].

2.3.1.1 Process description of delayed coking

A schematic flow diagram of the delayed coking is shown in **Figure 1**. The process includes a fractionator, furnace, two coke drums, and stripper. the feedstock is charged directly to the fractionator, where it is heated, and the lighter fractions are removed as middle distillates. The bottom of the fractionator is pumped to the coking furnace and then heated to the temperature range of 485–500°C. The heated feedstock (liquid—vapour mixture) enters one of the pairs of coking drums, where the cracking reactions continue. The energy obtained in the furnace passages is sufficient to perform the cracking reaction when the coking drum is filled. In the furnace, steam is injected to prevent the formation of premature coking. In addition, to prevent the formation of coke in the furnace, short residence time and high mass velocity in the furnace are required. Overhead stream in the coking drum; gases, naphtha, middle distillates and coker heavy gas oil are sent to the fractionator for separation, then separated and sent to downstream units for post-treatment and coke deposits on the inner surface. For continuous operation, two coke drums are used; while one is onstream, the other is decoking. The typical volume of a modern coke drum is about 1000 m^3, with a size range of 5–9 meters in diameter and a height range of 20–45 meters. The temperature in the coke drum ranges from 415 to 465°C and the pressure varies between 2 and 6 bar. Coker heavy gas oil is recycled as a coker feed and combined with fresh preheated feed and fed to the furnace, or used in other refining processes such as hydrocracker or gas oil hydrotreater or as a catalytic fluid cracking feed. The Coke drum is usually onstream about 24 hours before filling with porous coke. **Figure 2** shows a section of a coke drum and shows how coke forms during a delayed coking operation. The material at the bottom of the coke drum is fully carbonised, creating a porous structure through which gases and liquids can pass. The top layer is not fully carbonised until it is exposed to heat for a long time. Some foam forms on the top of the drum, so foam forming can be prevented by injecting anti-foam materials (silicone oil) into the coke drums during the last 5 or 6 hours of the coking cycle. It is important to prevent the carryover of foam into vapour lines. Level

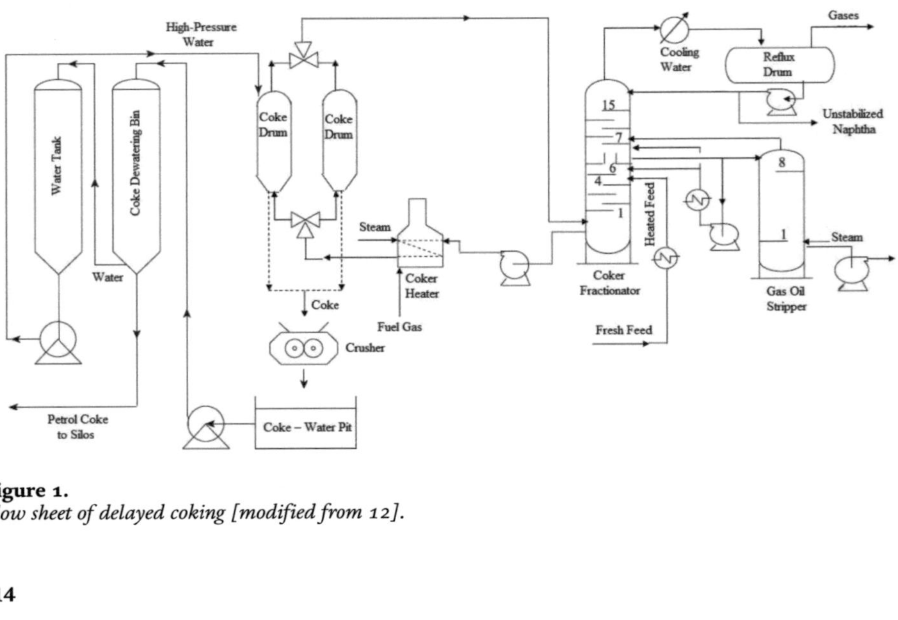

Figure 1.
Flow sheet of delayed coking [modified from 12].

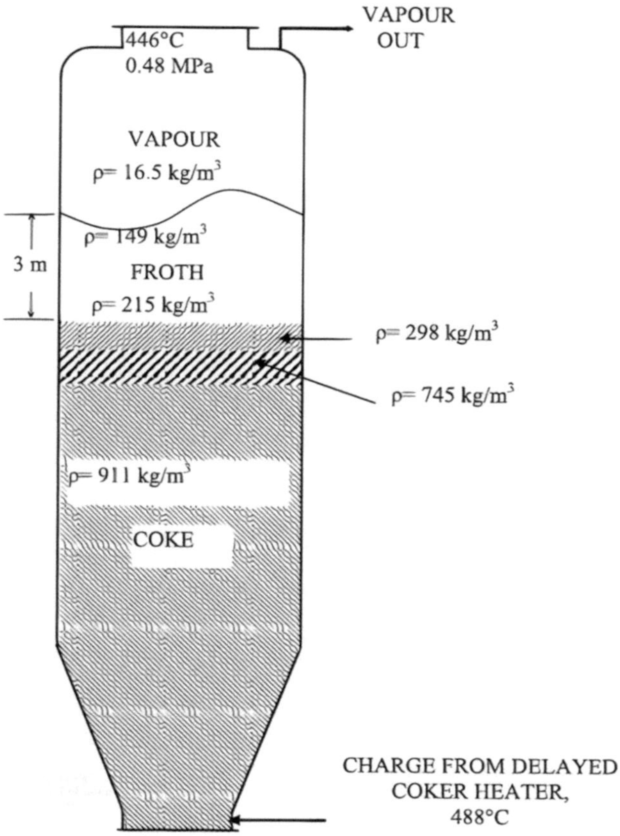

Figure 2.
Coke formation in coke drum of a delayed coking unit [6].

indicators are useful for detecting the position of liquid or foam in the drum. After steaming and cooling the coke drum, the coke is removed by drilling and cutting with high-pressure (up to 340 bar) water jets [6, 12–17].

Decoking operation of the drum (**Figure 3**) involves the following steps:

1. The coke deposit is cooled with water.

2. One of the heads of the coking drum is removed to make it possible to drilling of a hole through the deposit centre.

3. A hydraulic cutting machine, which uses multiple high-pressure water jets, is inserted into the hole and wet coke is removed from the drum [6].

Most cokers were originally designed for a 20- to 24-hour coking cycle. In the late 1980s and early 1990s, the coking cycle time was reduced to 16–20 hours. In the late 1990s, it dropped to 14 hours. A typical time cycle in delayed coking is shown in **Table 1**.

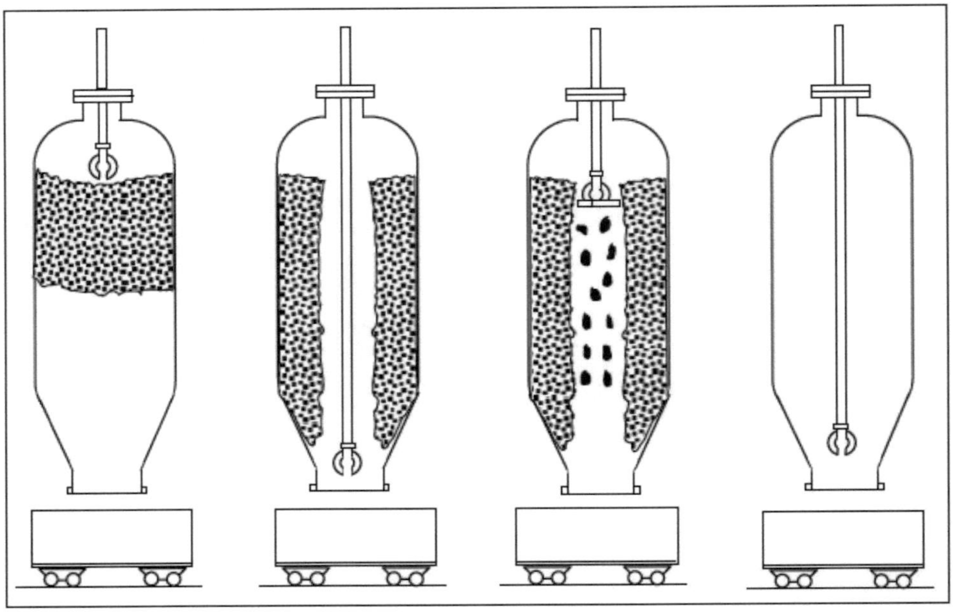

Figure 3.
Steps of decoking operation [14].

Operation	Time, h
Coking	24
Decoking	24
Switching drums and steaming out	3
Cooling with water	3
Draining water	2
Hydraulic decoking	5
Reheading and testing	2
Warming up	7
Spare time	2
Total cycle time	48

Table 1.
Time cycle for delayed coking [6, 14].

2.3.1.2 Delayed coking process variables

Delayed coking process variables include process operating variables, feedstock properties and engineering variables. Furnace outlet temperature, coke drum pressure and recycle ratios are the main operating variables that affect not only the coke yield but also its properties. Increasing the drum pressure leads to a higher coke yield and a slight increase in gas yield, because more molecules, even in the gas oil range, contribute to coke formation by remaining in the liquid phase. It also reduces the sulphur content of coke. However, refinery economics requires operating at minimal coke formation.

Variable	Effect on	
	Coke yield	Coke quality
Increase drum pressure	Increase	Variable
Increase drum temperature	Decrease	Improve
Increase coker recycle ratio	Increase	Improve to maximum
Thermal crack recycle	Increase	Improve

Table 2.
Effect of operating variables on the yield and quality of coke [6].

As the temperature of the furnace and drum increases, due to the removal of more volatile matter, the yield of coke reduces and the higher quality and harder coke is produced. However, it can cause cutting problems during decoking. Lower temperatures produce more coke, but lower quality. Therefore, the temperature at the furnace outlet must be optimized to form a minimum amount of coke in the furnace coils. To reduce the formation of coke in furnace coils, steam is injected into the furnace before the critical decomposition zone. However, the coke produced by steam injection in this process is more isotropic, that is, of lower quality. The recycle ratio has the same pressure effect as in delayed coking units, which varies from 1.03 to 1.30. The highest values are used in commercial units that produce premium coke, while the lowest values are used in delayed coking units where the goal is to maximise distillate yields. In addition, reducing the recycle ratio causes low-quality coke because the concentration of asphaltenes in the reaction mixture is higher [6, 12–17].

Delayed coking units for processing vacuum residues are designed to operate under operating conditions that maximise liquid distillates yield and minimise coke production. These operating conditions include lower pressures, higher temperatures, and a lower recycle ratio. Feedstock variables are characterization factors and conradson carbon that affect product yields. Engineering variables also affect process performance, including mode of operation, capacity, and equipment used in coking and handling equipment. Operating variables have practical constraints that prevent further changes. Also, the constraints for each will be different with the type of feed consumed [14]. The effect of operating variables on coke yield and quality is shown in **Table 2**.

2.3.1.3 Delayed coking feedstock

The delayed coking process can be applied to all residues in general, as they are not limited to constraints such as metal, sulphur, and asphaltene content. Heavy residues such as atmospheric and vacuum residue usually enter the delayed cokers, however, there are many raw materials that have been used as delayed coker feedstock for years. These feedstocks include:

1 - Gilsonite.
2 - Lignite pitch.
3 - Crack components (visbroken tar, cycle oil, decant oil or thermal crack tar).
4 - Refinery hazardous wastes.
5 - Deasphalted residues (pitch).
6 - Coal oils.
7 - Used plastic materials (recycling).

8 - Topped bitumen.
9 - C_3 to C_6 asphalt or lube oil extracts [6, 12, 15, 18]

2.3.1.4 Delayed coker yield prediction

In general, the products of the delayed coking process (based on vacuum residue feed) include gas (approximately 13 wt. %), naphtha (approximately 11 wt. %), middle distillate (approximately 45 wt. %), and green petroleum coke (approximately 31 wt. %).

The yield of products from delayed coking depends on the feed composition, in particular the amount of micro carbon residue (MCR) or Conradson carbon residue (CCR) content. Product yields can be estimated using the correlation based on the weight percentage of Conredson carbon residue (wt. % CCR) in the vacuum residue [14].

$$Gas\left(C_4^-\right)\ wt.\% = 7.8 + 0.144^*\ (wt.\%CCR) \tag{1}$$

$$Naphtha\ wt.\% = 11.29 + 0.343^*\ (wt.\%CCR) \tag{2}$$

$$Light\ Naphtha\ wt.\% = 0.3322^*\ (Naphtha\ wt\%) \tag{3}$$

$$Heavy\ Naphtha\ wt.\% = 0.6678^*\ (Naphtha\ wt\%) \tag{4}$$

$$Coke\ wt.\% = 1.6^*\ (wt.\%CCR) \tag{5}$$

$$Gas\ Oil\ wt.\% = 100 - (Gas\ wt.\% + Naphtha\ wt.\% + Coke\ wt.\%) \tag{6}$$

$$Light\ Cycle\ Gas\ Oil\ wt.\% = 0.645^*\ (Gas\ Oil\ wt.\%) \tag{7}$$

$$Heavy\ Cycle\ Gas\ Oil\ wt.\% = 0.355^*\ (Gas\ Oil\ wt.\%) \tag{8}$$

The gaseous compounds from the delayed coking process typically include methane, ethane, propane, butane, carbon monoxide, carbon dioxide, hydrogen, nitrogen, hydrogen sulphide and ammonia, the composition of which depends on the type of feed and the operating conditions.

2.3.1.5 Types of coke and their properties

Depending on the properties of feedstock and the operating conditions of the delayed coking process, different types of the coke can be produced. Coke can be distinguished by its morphology. Typically, coke can be divided into spherical shot coke (isotropic, amorphous, with almost no pores), sponge coke (semi-isotropic), and needle coke (anisotropic, regular crystalline structure, containing numerous fine pores and crystal sizes in the order of 4–7 nm). Either, according to its use, can be divided to fuel grade coke (cement industry and power generation), anode grade coke (aluminium production) or electrode grade coke (steel production). The differences between these types of coke are not always very clear. Due to the heterogeneity within the coke drum, one coke type may contain certain values of another coke type. Therefore, sponge coke may contain some shot coke and needle coke may contain some sponge coke [6, 15, 19]. Types of coke resulting from the delayed coking process with their optical structure are shown in **Figure 4**.

Petroleum coke can be in two forms, green petroleum coke and calcined petroleum coke. Petroleum coke obtained without calcination is called green coke. Coke calcination is done in a furnace to remove remaining hydrocarbons by heating green coke to about

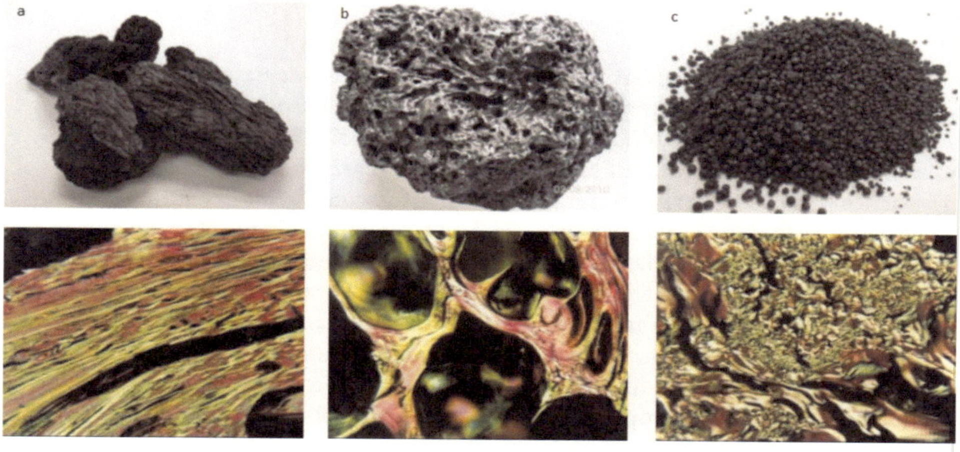

Figure 4.
Delayed coke types and optical textures. a: Needle coke, b: sponge coke, c: shot coke [20].

Properties	Fuel-grade green coke	Anode-grade calcined coke	Calcined needle coke
Sulphur (wt. %)	3–7.5	1.7–3.5	<0.5
Ash (wt. %)	0.1–0.3	0.1–0.4	<0.5
Nickel (ppm)	—	165–200	7
Vanadium (ppm)	200–400	120–350	—
Volatile matter (wt. %)	14 maximum	0.5	0.5
Bulk density (g/cm^3)	—	0.87	—
Real density (g/cm^3)	—	2.05	2.1–2.14

Table 3.
Typical properties for different types of coke [12].

1300–1500°C. During calcination, the coke decomposes further, and the carbon to hydrogen ratio increases from about 20 in green coke to 1000 for calcined coke [18].

Typical properties for different types of coke are shown in **Table 3**:

2.3.1.5.1 Shot coke

Shot coke comprises dense low porosity spherical clusters with 2–10 mm diameters, frequently present as agglomerates up to the size of basketballs. These large agglomerates are fragile and can be broken easily; however, the small spheres are very hard. Shot coke is obtained from petroleum precursors with high resin and asphaltene and low API gravity, and it is less valuable than sponge coke. High velocities in the reactor are required to produce shot coke with spherical particles. Given that a very turbulent condition is required for the formation of shot coke, shot-coke production in the laboratory is difficult, because surface velocities are very low [14, 19].

2.3.1.5.1.1 Variables affecting shot-coke formation

The variables which impact coke structure are the quality of the feedstock and the operating variables including pressure, temperature, vapor velocity, and recycle ratio.

- Feedstock quality:

Different authors agree that the feedstock properties associated with the production of shot coke are asphaltene content and Conradson carbon residue content. Researchers claim that the tendency to produce shot coke increases when the ratio between the asphaltene content and the Conradson carbon residue content approaches 0.5. Moreover, the characterisation of vacuum residues from different heavy oil sources shows that this ratio (asphaltene content/Conradson carbon content) is equal to or higher than 0.5; therefore, if the operating conditions are favourable, the formation of shot coke is likely when these feedstocks are processed.

Another fact that shows that the feedstock quality has an important impact on the coke structure is the use of decanted oil mixed with vacuum residue. Decanted oil is the residual product from the fluid catalytic cracking (FCC) process. This hydrocarbon stream is highly aromatic (more than 70% aromatics) and its incorporation into the coker with the feedstock (between 15% and 20% of the total feedstock) suppresses shot-coke formation. This suppressing action can be related to the solubility effect of the aromatics on the asphaltenes, although, this has not been shown experimentally [6, 14, 19].

- Operating variables

Operating variables refer to the pressure, temperature, vapour velocity, and recycle ratio within the coker.

Pressure: Reduction of the coker pressure favours the formation of shot coke.

Temperature: Higher temperatures favour shot-coke formation, and temperature change of 5°C or less can either suppress or promote shot-coke formation. In a commercial delayed coking unit, the heater outlet temperature varies between 490 and 500°C. However, scaling down of these units is reached by operating the small-scale units at lower temperatures, which may vary between 417 and 450°C.

Vapour Velocity: The feedstock flow is not an important variable that affects product yields in delayed coking technology, but this variable is an important parameter for shot-coke formation because it impacts the vapour superficial velocity, which is thought to give a spherical shape to shot-coke particles. The vapour superficial velocities in commercial delayed coking units are between 0.12 and 0.21 m/s. These vapour velocities are so high that they are not achieved in laboratory-scale units.

Recycle Ratio: It is calculated with the following expression:

$$RR = \mathrm{HF}/\mathrm{FF} \tag{9}$$

HF is the flow of the heater. After mixing the recycling flow with fresh feed at the bottom of the main fractionator, it is measured at the heater inlet. FF is the fresh feed stream that is measured before pumping the processed feedstock into the main fractionator. Both flows are measured in barrels per day.

The recycle ratio in delayed coking units varies from 1.03 to 1.30. The highest values are used in commercial units that produce needle coke, while the lowest values are used in delayed coke units where coke yields should be minimised [6, 14, 15, 18, 19].

2.3.1.5.2 Sponge coke

Sponge coke is the most common form of green coke. Sponge coke is a friable solid material with pores on the surface and internal cavities connecting the pores, which is due to the evolution of gas from the liquid in the coke drum. The structure of this coke causes good drainage of water from the coke drums and easy cutting of the coke bed with water jets. This coke is typically derived from crude oil, which contains numerous cross-linkages. The diffusion of gas bubbles into the coke drum may also cause some spongy coke. In fact, sponge coke is a combination of sponge and shot structures. Most sponge coke is used to fuel boilers. Some low-sulphur, low-metal sponge coke can be used to make anodes used in aluminium production [6, 14].

2.3.1.5.3 Needle coke

Using the proper feedstocks, optimal design techniques, and operating parameters, delayed coking can be used to produce needle coke, a specialized and rare product in the refining and coke production industry.

Producing good quality needle coke is not easy, because the control of several parameters is necessary to control the production process. In other words, it is a control process of several parameters. Needle coke is a premium coke made from special petroleum feedstocks. The needle coke has a silvery-grey appearance that has a broken crystalline needle-like structure, highly ordered, microcrystalline, under a light microscope. The observed optical texture is called flow domain. Needle coke has anisotropic components such as fine fibrous and leaflet structure. This coke has long, thin cavities that result from the gas bubbles released by the solid coke itself. This high-quality coke can only be produced from feedstocks of high purity (low metals and sulphur) and with high aromatic compounds, such as cycle oil from the fluid catalytic cracking unit. In addition, a long filling time is required for the solid coke in the coke drum to react and release the gases. This type of coke cannot be produced from vacuum residue [6, 14, 15, 21].

2.3.1.5.3.1 Needle coke applications

Natural graphite is a limited source. It is estimated that 800 million tons can be mined worldwide. Only 10 to 15% of natural graphite is actually graphite carbon. Most of it is amorphous and contains minerals or silicate metals. In contrast, needle coke is continuously produced with high graphitizable content and low impurity concentration [12].

It was generally accepted that needle coke can be divided into two types according to the different feedstocks and named coal-based needle coke and petroleum-based needle coke. Excellent physical and chemical properties of needle coke such as high mechanical strength, high electrical conductivity (strong oxidation resistance), high thermal conductivity, high density as well as low thermal expansion coefficient (good abrasion resistance/heat shock resistance), low ash and sulphur content, low volatility, low energy consumption and easy graphitizable make needle coke an excellent raw material to obtain high-quality artificial graphite [12, 22].

There are two methods, basic oxygen furnace (BOF) and electric arc furnace (EAF), for steel production. Coal, iron, and limestone are used to produce steel in the BOF method. However, in the EAF method, an electric current passes through the graphite electrodes to convert the steel scrap into molten steel. Approximately 70% of world steel is produced by the BOF method and 30% by the EAF method. EAF has historically been the fastest growing sector of the global steel industry, with EAF steel production amounting to about 20 million tonnes per year in 1950, and EAF steel

Property	Quality grade		
	Super premium	Premium	Intermediate
Coefficient of thermal expansion (CTE), $*10^{-7}$/°C	<2.0	2.0–3.0	3.1–4.0
Sulphur content, wt.%	<0.5	<0.6	<0.8
Real density, gm/cc	>2.12	>2.12	>2.12
Ash content, wt.%	<0.1	<0.2	<0.2
H content, wt.%	<0.03	0.03–0.05	0.03–0.05

Table 4.
Typical calcined needle coke specification [6, 12].

production expanded rapidly after 1950, and it exceeded 100 million tons in the 1970s. Needle coke, produced in the delayed coking process of petroleum oil refineries, was later developed in 1960 and commercialised in 1970. Finally, EAF steel production in 2020 reached about 550 million tons [12].

Inputs/initial costs of steel production through the EAF method include scrap steel, electricity, and graphite electrodes. There is no known alternative to graphite electrodes used in the EAF method of steel production. Needle coke is a major component in the production of graphite electrodes. The main application of needle coke is in the graphite electrode industry, and it can be purchased for 1500–3000 $/ton. In addition, needle coke is also used in the production of graphite cathodes in the aluminium industry. Electrodes made of needle coke need to withstand temperatures above 3000°C. Global steel production on the EAF is expected to grow. This has led to a similar increase in consumption of graphite electrodes. It is expected to eventually increase the consumption of needle coke [6, 12].

Needle coke is now widely used as a carbon filler for the production of graphite electrodes in the steel industry for smelting scrap metal for recycling in an electric arc furnace (EAF), cathodes required for smelting aluminium, anodes for commercial lithium-ion batteries, electric machines and some inherent parts of mobile phones, electrode materials for high energy density supercapacitors, anode materials for high-performance sodium-ion batteries, adsorbents, isotropic graphite, nuclear graphite, perovskite solar cell, carbon substitute super-activated carbon, graphene precursors, aerospace and other functional materials are used. Graphite electrodes have a low coefficient of thermal expansion (CTE), which is defined as an increase in length per unit temperature increase. Low CTE values indicate anisotropic needle coke, while high values indicate an isotropic shot coke [6, 12, 22, 23].

In terms of grade, needle coke is divided into an intermediate, premium, and super premium needle coke. As shown in **Table 4**, their difference is in the amount of thermal expansion coefficient and sulphur content.

2.3.1.5.3.2 Feedstocks quality for needle coke production

Precursors for needle coke production have historically been limited to available residues whose aromatic molecular composition naturally predisposes them to form highly anisotropic carbon during carbonisation. However, further requirements of the feedstock include:

- Low ash content
- Low quinoline insoluble (QI) content
- Low asphaltene content
- Reduced content of stable nitrogen or sulphur heterocyclics
- Low oxygenate content
- Low air and carboxy reactivity of the coke during calcination

Coal-based needle coke is made from Coal Tar Pitch, refined coal tar pitch, refined coal liquefied pitch, and coal extraction. Petroleum-based needle coke is usually obtained by delayed coking of residual oil, petroleum bitumen, oxidized petroleum bitumen, and Fluidised Catalytic Cracker Decant Oil [18].

The chemical and physical properties considered in choosing a proper feedstock for the production of needle coke are summarised as follows:

1. Feedstock should have high aromaticity with 60–85% aromatic carbon aromaticity;

2. Feedstock should be of high initial boiling point, over 250°C with not more than 25–30% of material boiling below 360°C;

3. Feedstock should have low API gravity;

4. Feedstock should have low sulphur content preferably below 1 wt. % due to the concern for product quality;

5. Feedstock should have low metal, asphaltenes, and CCR content [6].

2.3.2 Fluid coking

Although the delayed coking process has been selected for large-scale operations, they are more attractive for processing the small volumes of residues due to the safety issues involved in decoking the drums at the end of each cycle. In addition, by reducing the retention time of cracked vapours, the yields of coking distillation products can be improved. To simplify the handling of the coke and to enhance product yields, Exxon developed a continuous process in the mid-1950s called fluidized bed coking (or fluid coking), in which the residence time was shorter, with more liquid and less coke. However, in this process, the products have lower quality. Fluid coking is a fluidized bed process developed by fluid catalytic cracking (FCC) technology, except that no catalysts are used and heavy feedstocks such as atmospheric and vacuum residues, residues of catalytic cracking units and oil sand bitumen turn into light products. In fluid coking, about 6% of the coke is burned to provide heat to the process, while the net coke yield is 70 to 75% of delayed coking. The yields of products resulting from fluid coking are determined by feed properties, fluidized bed temperature, and residence time in the bed [12, 14–17].

Feed	Products	Yield wt.%
Arab light Vaccum residue (22 wt.% CCR)	Reactor Gas, C_4^-	11 wt.%
	Coker Naphtha (C_5–221°C)	15–20 wt.%
	LCGO[1] (221–343°C)	12–14 wt.%
	HCGO[2] (343–524°C)	35–36 wt.%
	Fuel gas	0.02 FOEB[3] bbl Feed
	Net coke	21 wt.%

[1]Light Coker Gas Oil.
[2]Heavy Coker Gas Oil.
[3]Fuel Oil Equivalent Barrels.

Table 5.
Yield of fluid coker process [14].

An example of the material balance for fluid coking of Arab light vacuum residue is given in **Table 5**.

2.3.2.1 Process description of fluid coking

Fluid coking is a thermal cracking process consisting of a fluidized bed reactor and a fluidized bed burner. A flow diagram is shown in **Figure 5**. Vacuum residue is preheated and fed to a scrubber that operates at 370°C above the reactor for coke fine particle recovery. The heavy hydrocarbons in the feed are recycled with the fine particles to the reactor as slurry recycle. The heavy vacuum residue feed is injected through nozzles to a fluidized bed of coke particles. Cracking reactions take place in the reactor at a temperature of 500–550°C, and the feed is converted to vapour and lighter gases, which enter the scrubber after passing through the cyclones at the top of the reactor and go to the fractionator column. Steam enters from the bottom of the reactor to remove heavy hydrocarbons from the coke surface. The evolution of vapour from the cracking of the feed, and the addition of steam, gives intense mixing of the

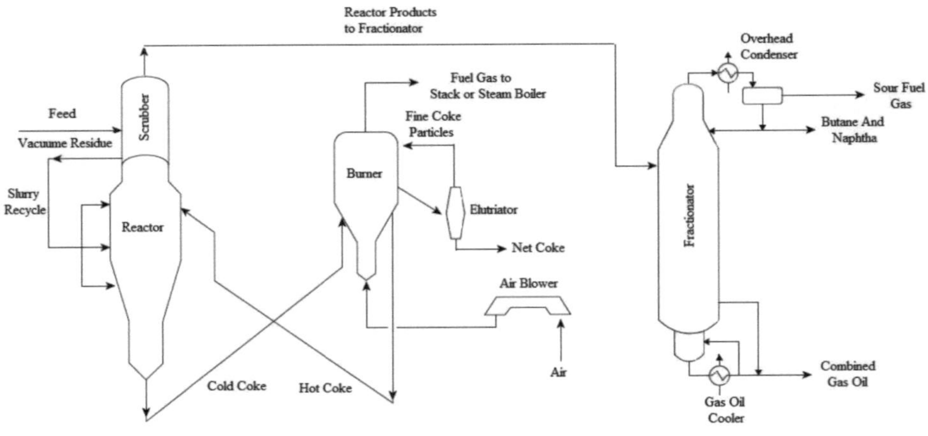

Figure 5.
Flow sheet of fluid coking [modified from 12].

coke particles within the reactor. The coke formed in the reactor flows continuously to the burner, where it is heated to 593–677°C and burns with partial combustion of 15–30% of the coke by injecting air into the burner. Coke combustion produces flue gases with low heating value (20 BTU/SCF), which are rich in CO and H_2. Parts of the heated coke particles are returned to the reactor to provide energy for the endothermic cracking reactions and to maintain the reactor temperature. After cooling, the remaining coke is removed from the process as a stream of fine particles of 'petroleum coke' and is burned in power plants or cement industries. This coke is very isotropic, rich in ash and sulphur and therefore not used in the carbon and graphite industry [12, 16, 17].

The lower limit on operating temperature for fluid coking is set by the behaviour of the fluidized coke particles. If the conversion to coke and light ends is too slow, then the coke particles become sticky and agglomerate within the reactor. This phenomenon occurs in localised zones of the reactor, likely near the nozzles that inject the (colder) liquid bitumen feed, giving rise to chunks of coke that fall to the bottom of the bed. For this reason, optimising the method for introducing feed into the reactor is crucial. In addition, excellent heat transfer in the fluidized bed helps to reduce hotspots, which allows the reactor to operate at a higher temperature to cause more cracking of volatile matters. These factors generally reduce coke yields and increase the yields of gas oil and olefins compared to the delayed coking process. One disadvantage of the fluid coking process is the high rate of coke accumulation inside the unit. The reactor operates in a fouling mode, so coke deposits continuously on the interior surfaces during operation. The reactor must be shut down for a month or more every 2 or 3 years to remove the accumulated coke, which can grow to be as thick as 1 meter on the interior walls of the coker. The second disadvantage is the emission of significant amounts of hydrogen sulphide and sulphur dioxide from the reactor burner [16, 17].

At first, it was thought that the fluid coking process would replace the delayed coking process in the market, but so far this has not happened.

2.3.3 Flexicoking

The decline in coke markets derived from delayed coking and fluid coking due to constraints in sulphur emissions encouraged the development of flexicoking. Burning coke to generate process heat (**Figure 6**) liberates the sulphur in the coke as hydrogen sulphide and sulphur dioxide gases. The off-gas stream from the coke burner also contains CO, CO_2 and N_2. An alternate approach is to use a coke gasifier which can convert the carbonaceous solids to a mixture of CO, CO_2 and H_2 without producing SO_2. Flexicoking was designed by ExxonMobil as a fluid coking modifier that was introduced in 1976 in Japan. This process combines fluid coking with coke gasification, which, similar to fluid coking, is a fluidized bed process developed from catalytic fluid cracking technology. A fluidized bed is added to the process, which acts as a gasifier in which coke from the heater is reacted with steam and air in a fluid-bed gasifier to produce a gas of low heating value (20–40 BTU/sCF) and significantly reduces coke production. Yields of liquid products are the same for flexicoking and fluid coking because the coking reactor is unaltered, but up to 97% of the coke can be converted to gas by steam and air in a gasifier. Air is injected into the gasifier to maintain temperatures of 830–1000°C, but injected air is not enough to burn the entire coke. Under these conditions, the sulphur in the coke is converted to hydrogen sulphide, which can be scrubbed from the gas prior to combustion elsewhere.

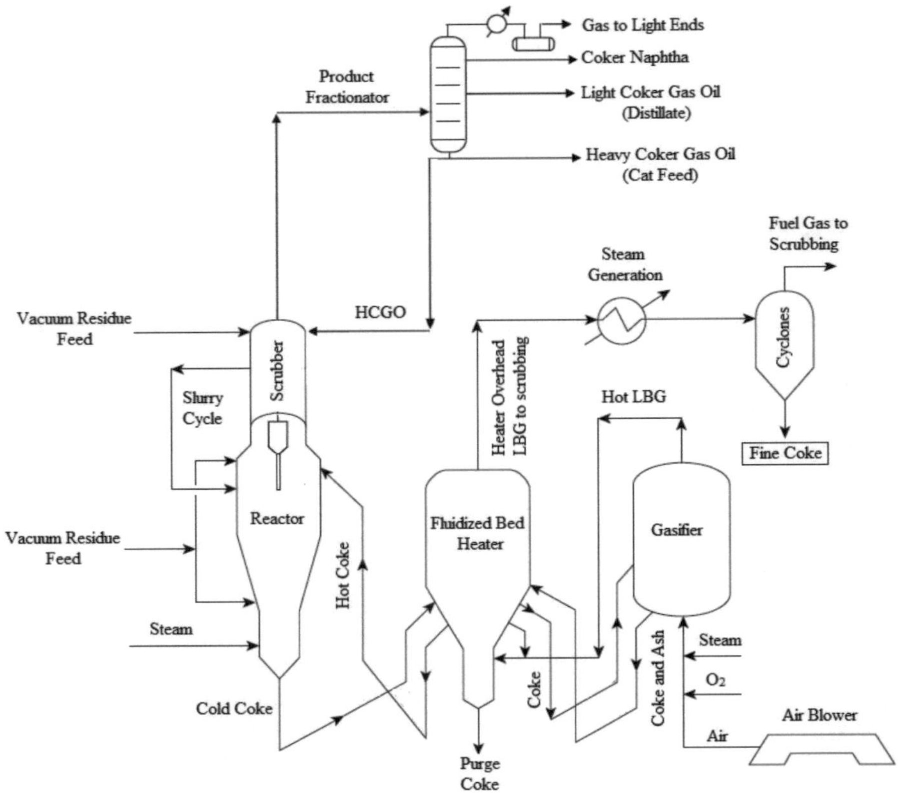

Figure 6.
Flow sheet of flexicoking [modified from 12].

After removal of the hydrogen sulfide, a typical gas product contains 18% CO, 10% CO_2, 15% H_2, 51% N_2, 5% H_2O and 1% CH_4. Petroleum coke is removed, and economical fuel gas is available for use at the refinery. Due to the high initial investment and mechanical cost, only seven units were built worldwide. The main drawback of gasification is the requirement for a large additional reactor, especially if the high conversion of the coke is required [12, 14–17].

2.3.3.1 Process description of flexicoking

In the process, the viscous feedstock enters the scrubber for direct-contact heat exchange with the overhead product vapours from the reactor. Lower-boiling overhead constituents in the scrubber go to a conventional fractionator and also to light ends recovery. The feedstock is thermally cracked in the reactor fluidized bed to a range of gas and liquid products and coke. The typical bed temperature is 510–540°C. Vapour products resulting from the conversion reactions in the bed pass through the cyclone separators, which remove most of the entrained coke and return it to the reactor bed. The cyclone outlets discharge the vapor product directly into a scrubber, where the heavy liquid is used to scrub out the remaining coke dust and condense unconverted high-boiling fractions. The dust-laden liquid is recycled as 'a slurry cycle' to the reactor with the feed. The scrubbed vapour is sent to the coker fractionator, where the stream is split into gas, naphtha, distillate and heavy gas oil streams.

The heater is located between the reactor and the gasifier, and it serves to transfer heat between the two vessels. The heater temperature is controlled by the rate of coke circulation between the heater and the gasifier. Adjusting the air rate to the gasifier controls the unit inventory of coke, and the gasifier temperature is controlled by steam injection into the gasifier. Excess coke is converted to a low-heating value gas in a fluid-bed gasifier with steam and air. The air is supplied to the gasifier to maintain temperatures of 830–1000°C, but is insufficient to burn all the coke. The heater transfers heat from the gasifier overhead gas to coke, which in turn supplies the heat of reaction in the reactor. The heater bed temperature is approximately 610°C. Coke is continuously circulated between the three vessels to transfer heat and maintain vessel inventories. A typical gas product, after the removal of hydrogen sulfide, contains carbon monoxide (CO, 18%), carbon dioxide (CO_2, 10%), hydrogen (H_2, 15%), nitrogen (N_2, 51%), water (H_2O, 5%) and methane (CH_4, 1%) [12, 14–17].

In the oxidation zone of the gasifier, the following reactions take place very rapidly [14]:

$$C + 0.5O_2 \rightarrow CO \tag{10}$$

$$CO + 0.5O_2 \rightarrow CO_2 \tag{11}$$

In the reduction zone, the following reactions take place slowly:

$$C + H_2O \rightarrow CO + H_2 \tag{12}$$

$$H_2O + CO \rightarrow CO_2 + H_2 \tag{13}$$

Delayed coking is the most commonly used process among all commercial coking processes. More than 92% of petroleum coke is produced in the delayed coking process; About one-third of feed streams are produced in the form of petroleum coke. Due to the reaction conditions, net coke production from fluid cokers and flexicokers is only about 5–10 wt.% of the feed material. About 20–25% of 700 refineries worldwide are equipped with delayed cokers. Of the 140 US refineries in operation, 55 have delayed coker units. Most of the petroleum coke is produced in the United States, followed by China, South America, Canada, India, the Middle East and Western [6, 12].

Coke produced by delayed coker is a marketable product, while coke produced by fluid coker and flexicoker is burned to meet the reactor heat needs and feed preheat.

3. Conclusion

At present, light crude oil reserves are the main source of energy that meets global energy demand due to high quality and low production costs. Decline in light crude oil reserves poses great challenges to meeting the world's energy needs. Heavy oil and oil residues have become a suitable alternative source to meet global energy demand. According to the approaches to achieving higher H/C ratios, upgrading technologies can be divided into carbon rejection and hydrogen addition processes. However, the cost of hydrogen addition processes is much higher than carbon rejection processes, because the production of hydrogen and the catalysts used in hydrogen addition processes are very expensive. Carbon rejection technologies have been used by refineries since 1913 to upgrade various hydrocarbon feeds. In a carbon rejection process, raw materials are heated to high temperatures to crack large hydrocarbons into smaller ones. Coking (delayed, fluid and flexi) is one of the types of carbon rejection processes. Delayed coking has been chosen by many refineries as an upgrading process due to its low investment cost and the inherent flexibility of the process to process any residuals. In this process, 20–30 wt.% coke is produced as a by-product. Depending on the properties of the raw materials and the operating conditions of the delayed coking process, different types of the coke can be produced. Typically, coke can be divided into spherical shot coke, sponge coke, and needle coke. Using the proper feedstocks, optimal design techniques, and operating parameters, delayed coking can be used to produce needle coke, a specialized and rare product in the refining and coke production industry. Needle coke is a premium coke made from special petroleum feedstocks. There are two methods, BOF and EAF, for steel production. Coal, iron, and limestone are used to produce steel in the BOF method. However, in the EAF method, an electric current passes through the graphite electrodes to convert the steel scrap into molten steel. There is no known alternative to graphite electrodes used in the EAF method of steel production. Needle coke is a major component in the production of graphite electrodes. The main application of needle coke is in the graphite electrode industry. Global steel production on the EAF is expected to grow. This has led to a similar increase in consumption of graphite electrodes. It is expected to eventually increase the consumption of needle coke.

Author details

Jafar Ramezanzadeh* and Hossein Moradi
University of Tehran, Catalysis and Nanostructured Materials Research Laboratory, School of Chemical Engineering, College of Engineering, Tehran, Iran

*Address all correspondence to: jafarramezanzadeh@ut.ac.ir

References

[1] Bagheri SR. Mesophase Formation in Heavy Oil [thesis]. Edmonton: University of Alberta; 2012

[2] Sahu R, Song BJ, Im JS. A review of recent advances in catalytic hydrocracking of heavy residues. Journal of Industrial and Engineering Chemistry. 2015;**27**:12-24. DOI: 10.1016/j.jiec.2015.01.011

[3] Hashemi R, Nassar NN, Almao PP. Nanoparticle technology for heavy oil in-situ upgrading and recovery enhancement: Opportunities and challenges. Applied Energy. 2014;**133**: 374-387. DOI: 10.1016/j.apenergy.2014.07.069

[4] Nguyen MT, Nguyen NT, Cho J. A review on the oil-soluble dispersed catalyst for slurry-phase hydrocracking of heavy oil. Journal of Industrial and Engineering Chemistry. 2016;**43**:1-12. DOI: 10.1016/j.jiec.2016.07.057

[5] Eshraghian A. Thermal and Catalytic Cracking of Athabasca VR and Bitumen [thesis]. Calgary: University of Calgary; 2017

[6] Sawarkar AN, Pandit AB, Samant SD. Petroleum residue upgrading via delayed coking. The Canadian Journal of Chemical Engineering. 2007;**85**:1-24. DOI: 10.1002/cjce.5450850101

[7] DSJ J, Pujadó PP, editors. Handbook of Petroleum Processing. 1st ed. Dordrecht: Springer; 2006. p. 1353. DOI: 10.1007/1-4020-2820-2

[8] Rana MS, Sámano V, Ancheyta J. A review of recent advances on process technologies for upgrading of heavy oils and residua. Fuel. 2007;**86**: 1216-1231. DOI: 10.1016/j.fuel.2006.08.004

[9] Castañeda LC, Muñoz JAD, Ancheyta J. Combined process schemes for upgrading of heavy petroleum. Fuel. 2012;**100**:110-127. DOI: 10.1016/j.fuel.2012.02.022

[10] Prajapati R, Kohli K, Maity SK. Slurry phase hydrocracking of heavy oil and residue to produce lighter fuels: An experimental review. Fuel. 2021;**288**: 119686. DOI: 10.1016/j.fuel.2020.119686

[11] Tsujino T. Recent development of Texaco gasification technology and its applications. Fuel and Energy Abstracts. 1996;**37**:183. DOI: 10.1016/0140-6701(96)88538-9

[12] Jaeger H, Frohs W, editors. Industrial Carbon and Graphite Materials: Raw Materials, Production and Application. 1st ed. Weinheim: Wiley; 2021. p. 1008. ISBN: 9783527336036

[13] Gray MR. Upgrading Oilsands Bitumen and Heavy Oil. 1st ed. Edmonton: The University of Alberta Press; 2015. p. 499. ISBN: 9781772120356

[14] Fahim MA, Al-Sahhaf TA, Elkilani AS. Fundamentals of Petroleum Refining. 1st ed. Oxford: Elsevier; 2010. p. 516. ISBN: 9780444527851

[15] Huc AY. Heavy Crude Oils: From Geology to Upgrading: An Overview. 1st ed. Paris: IFP energies nouvelles publications; 2011. p. 516. ISBN: 9782710808909

[16] Speight JG. Heavy oil Recovery and Upgrading. 1st ed. Cambridge: Gulf professional publishing; 2019. p. 850. ISBN: 9780128130254

[17] Speight JG. Heavy Oil and Extra-Heavy Oil Upgrading Technologies.

1st ed. Oxford: Gulf professional publishing; 2013. p. 176. ISBN: 9780124045705

[18] Clark JG. The Modification of Waxy Oil for Preparing a Potential Feedstock for Needle Coke Production [thesis]. Pretoria: University of Pretoria; 2011

[19] Edwards L. The history and future challenges of calcined petroleum coke production and use in aluminum smelting. Journal of Metals. 2015; **67**:308-321. DOI: 10.1007/s11837-014-1248-9

[20] Vieman AA. The Impact of Phase Behaviour on Coke Formation in Delayed Cokers [thesis]. Toronto: University Toronto; 2002

[21] Escallón MM. Petroleum and Petroleum/Coal Blends as Feedstocks in Laboratory-Scale and Pilot-Scale Cokers to Obtain Carbons of Potentially High Value [thesis]. Pennsylvania: The Pennsylvania State University; 2008

[22] Zhu H, Zhu Y, Xu Y. Transformation of microstructure of coal-based and petroleum-based needle coke: Effects of calcination temperature. Asia-Pacific Journal of Chemical Engineering. 2021; **16**. DOI:10.1002/apj.2674

[23] Wang G. Molecular Composition of Needle Coke Feedstocks and Mesophase Development during Carbonization [thesis]. Pennsylvania: The Pennsylvania State University; 2005